VISION

Arising from the 2019 Darwin College Lectures, this book presents essays from seven prominent public intellectuals on the theme of vision. Each author examines this theme through the lens of their own particular area of expertise, making for a lively interdisciplinary volume including chapters on neuroscience, colour perception, biological evolution, astronomy, the future of technology, computer vision, and the visionary core of science.

Featuring contributions by professors of neuroscience Paul Fletcher and Anya Hurlbert, professor of zoology Dan-Eric Nilsson, the futurist Sophie Hackford, Microsoft distinguished scientist Andrew Blake, theoretical physicist and author Carlo Rovelli, and Dr Carolin Crawford, the Public Astronomer at the University of Cambridge, this volume will be of interest to anybody curious about how we see the world.

ANDREW FABIAN is an Emeritus Professor at the Institute of Astronomy of the University of Cambridge and Emeritus Fellow of Darwin College.

JANET GIBSON is the College Registrar for Darwin College at the University of Cambridge.

MIKE SHEPPARD is an Honorary Fellow of Darwin College, Cambridge.

SIMONE WEYAND is a Fellow of Darwin College, a Visiting Scientist at the Department of Biochemistry, and a Visiting Scientist at the Cambridge Institute of Medical Research.

THE DARWIN COLLEGE LECTURES

These essays are developed from the 2019 Darwin College Lecture Series. Now in their thirty-fourth year, these popular Cambridge talks take a single theme each year. Internationally distinguished scholars, skilled as popularizers, address the theme from the point of view of seven different arts and sciences disciplines.

Subjects covered in the series include

2019 VISION
eds. Andrew Fabian, Janet Gibson, Mike Sheppard and Simone Weyand
pb 9781108931021

2018 MIGRATION
eds. Johannes Knolle and James Poskett
pb 9781108746014

2017 EXTREMES
eds. Duncan J. Needham and Julius F. W. Weitzdörfer
pb 9781108457002

2016 GAMES
eds. David Blagden and Mark de Rond
pb 9781108447324

2015 DEVELOPMENT
eds. Torsten Krude and Sara T. Baker
pb 9781108447379

2014 PLAGUES
eds. Jonathan L. Heeney and Sven Friedemann
pb 9781316644768

2013 FORESIGHT
eds. Lawrence W. Sherman and David Allan Feller
pb 9781107512368

2012 LIFE
eds. William Brown and Andrew Fabian
pb 9781107612556

2011 BEAUTY
eds. Lauren Arrington, Zoe Leinhardt and Philip Dawid
pb 9781107693432

2010 RISK
eds. Layla Skinns, Michael Scott and Tony Cox
pb 9780521171977

Vision

Edited by *Andrew Fabian*

University of Cambridge

Janet Gibson

University of Cambridge

Mike Sheppard

University of Cambridge

Simone Weyand

University of Cambridge

CAMBRIDGE
UNIVERSITY PRESS

CAMBRIDGE
UNIVERSITY PRESS

University Printing House, Cambridge CB2 8BS, United Kingdom

One Liberty Plaza, 20th Floor, New York, NY 10006, USA

477 Williamstown Road, Port Melbourne, VIC 3207, Australia

314–321, 3rd Floor, Plot 3, Splendor Forum, Jasola District Centre, New Delhi – 110025, India

79 Anson Road, #06–04/06, Singapore 079906

Cambridge University Press is part of the University of Cambridge.

It furthers the University's mission by disseminating knowledge in the pursuit of education, learning, and research at the highest international levels of excellence.

www.cambridge.org
Information on this title: www.cambridge.org/9781108931021
DOI: 10.1017/9781108946339

First published 2021

Printed in Singapore by Markono Print Media Pte Ltd

A catalogue record for this publication is available from the British Library.

ISBN 978-1-108-93102-1 Paperback

This volume is dedicated to the memory of Professor Willy Brown CBE, who was Master of Darwin College from 2000 to 2012 and who died on 1 August 2019. Willy made lasting contributions to the Darwin College lecture series, and was much engaged with us in discussions of the Vision series. He was our close friend, and his memory continues to inspire us.

Contents

Figures

Notes on Contributors

Andrew Blake is a pioneer in the development of the theory and algorithms that make it possible for computers to behave as seeing machines. He trained in mathematics and electrical engineering in Cambridge UK and at MIT, and studied for a doctorate in Artificial Intelligence at the University of Edinburgh. He was Professor of Information Engineering at Oxford University and joined Microsoft in 1999 to found the Computer Vision group in Cambridge, before becoming Director of Microsoft's Cambridge Laboratory in 2010 and a Microsoft Distinguished Scientist. He is Chairman of Samsung's AI Research Centre SAIC in Cambridge and Scientific Adviser to the FiveAI autonomous driving company serving as an adviser to Siemens. In 2010, he was elected to the council of the Royal Society, and he was appointed to the board of the EPSRC in 2012. He was Director at The Alan Turing Institute 2015–2018. He has been Honorary Professor of Machine Intelligence at the University of Cambridge since 2007 and is a Fellow of Clare Hall. He has been a Fellow of the Royal Academy of Engineering since 1998 and Fellow of the Royal Society since 2005. He twice won the prize of the European Conference on Computer Vision, with R. Cipolla in 1992 and with M. Isard in 1996, and was awarded the IEEE David Marr Prize (jointly with K. Toyama) in 2001. The Royal Academy of Engineering awarded him their Silver Medal in 2006, and in 2007 he received the Institution of Engineering and Technology Mountbatten Medal. He was named a Distinguished Researcher in Computer Vision by the IEEE in 2009. In 2011, with colleagues at Microsoft Research, he received the Royal Academy of Engineering MacRobert Gold Medal for the machine learning at the heart of the Microsoft Kinect 3D camera. Exactly 80 years after Einstein, in 2014, he gave the Gibbs lecture at the Joint Mathematics Meetings. The BCS awarded him its Lovelace Medal and prize lecture in 2017. He holds honorary doctorates at the University of Edinburgh and the University of Sheffield.

Carolin Crawford is the Public Astronomer at the Institute of Astronomy at the University of Cambridge. She received her PhD from Cambridge University, and for many years she used X-ray, optical, and near-infrared observations to investigate the environments of some of the largest galaxies in the Universe. Her research was carried out alongside – and later eclipsed by – a growing role in the public communication of science. Carolin now gives many talks every year communicating the excitement of astronomy to as wide an audience as possible; she also makes regular appearances on local and national radio. Her efforts were recognised by a Woman of Outstanding Achievement award from the UK Resource Centre for Women in Science, Engineering and Technology for 'communication of SET with a contribution to society' in 2009, and her appointment as the Professor of Astronomy at Gresham College 2011–2015. Carolin is also a College Lecturer, Fellow, and Admissions Tutor at Emmanuel College, where she teaches mathematics.

Andrew Fabian is an Emeritus Professor at the Institute of Astronomy of the University of Cambridge and an Emeritus Fellow of Darwin College.

Janet Gibson is the College Registrar in Darwin College.

Paul Fletcher is Bernard Wolfe Professor of Health Neuroscience at the University of Cambridge, Director of Studies for Preclinical Medicine at Clare College, and Honorary Consultant Psychiatrist with the Cambridgeshire and Peterborough NHS Foundation Trust. He studied Medicine, before carrying out his specialist training in psychiatry and taking a PhD in cognitive neuroscience. He researches human perception, learning and decision-making, and is especially interested in hallucinations – perception in the absence of a stimulus – feeling that the existence of such phenomena offers us important insights into how our brains construct our experience of the world.

Sophie Hackford is a futurist whose research entails meeting weirdos and troublemakers in off-the-beaten-track labs, makerspaces, and garages around the globe – Shenzhen, Seoul, Detroit, Mumbai. As part of her research, she consults for exec teams and boards of large companies on understanding the explosive new technologies defining the new economy. Sophie is also CEO of a data and AI company, 1715 Labs, that she's currently spinning out of the Astrophysics department at Oxford University with her academic co-founder. This follows a career building businesses for WIRED magazine, for Singularity University at the NASA Research Park in Silicon Valley, and, prior to California, the interdisciplinary Oxford Martin School at Oxford University, where Sophie raised more than \$120 million of research investment.

Anya Hurlbert is Professor of Visual Neuroscience, Director of the Centre for Translational Systems Neuroscience, and Dean of Advancement at Newcastle University, where she co-founded and directed the Institute of Neuroscience. She trained as a physicist, physiologist, neuroscientist, and physician, at Princeton, Harvard, MIT, and Cambridge. Professor Hurlbert's research focuses on human vision; she lectures widely on colour perception and art, and has devised and co-curated several science-based art exhibitions, including an interactive installation in the 2014 exhibition Making Colour at the National Gallery, where she was Scientist Trustee.

Dan-Eric Nilsson is a Professor of functional zoology at Lund University in Sweden. He is a fellow of the Royal Swedish Academy of Sciences, and several other academic societies. He is the head of The Lund Vision Group, which is an internationally leading centre for comparative vision research. He co-authored the popular textbook *Animal Eyes* published by Oxford University Press.

Carlo Rovelli is a theoretical physicist, known for the development of loop quantum gravity, his work on the nature of physical time, and the relational interpretation of quantum mechanics. He was born in Italy in 1956, has studied in Bologna and Padova, and has worked in several universities in Italy and the United States; he is currently directing the Quantum Gravity group of the Centre de Physique Théorique of the University of Aix-Marseille. He has honorary degrees from Beijing Normal University and the Universidad Nacional de San Martín of Buenos Aires, Argentina. He is a member of the International Academy for the Philosophy of Science, of the Institut Universitaire de France, and of the Accademia Galileana. He has written successful popular science books: his *Seven Brief Lessons on Physics* has been translated into 41 languages and has sold over a million copies. His most recent book, *The Order of Time*, is on the nature of time.

Mike Sheppard is an Honorary Fellow of Darwin College.

Simone Weyand is a biochemist and biophysicist who has worked on the structure determination of membrane proteins, such as the bacterial transporter Mhp1 and the human histamine H1 receptor, by the use of X-ray crystallography. Dr Weyand was awarded a Sir Henry Dale Fellowship in 2013. Her Fellowship work uses a holistic approach to understanding the molecular mechanism of human neurotransmitter transporters by investigating the high-resolution structure and the functional analysis and trafficking in the cell. This combined approach, including different techniques, will provide deeper insights into the basic principle of action of these proteins and will eventually enable a more rational and efficient drug design. Dr Weyand is an Official Fellow of Darwin College.

Acknowledgements

Preparing and conducting the Darwin College Lecture Series on Vision and compiling this volume of essays would not have been possible without the help of many people. The editors would like to thank all of them for their help, which was very much appreciated. Professor Mary Fowler, Master of Darwin College, was much engaged throughout the genesis of this series, as were the members of Darwin's Education and Research Committee. We would like to thank Dr Sandy Skelton for her many contributions to preparing the lecture series, and Dr Tony Cox and Dr Iosifina Foskolou for their indefatigable help with running the individual lectures. We would also like to thank Espen Koht, who managed all things involving IT throughout the series and without whom we would have been all at sea.

Introduction

MIKE SHEPPARD

The 34th Darwin College Lecture Series, held in 2019, addressed Vision. The aim of these lectures, as with all the Darwin College Lectures, was to provide an interdisciplinary study. The lectures range widely: they survey the mechanisms of visual perception, and the evolution of eyes; they address the mental processes underpinning vision, and the nature and significance of private visions and hallucinations; they explore the vision and imagery of artists and of scientists in their endeavours to elucidate the world. The discussions encompass astronomical observation, which enables us to look back over the evolution of the Universe to the earliest epochs, and they extend to foresight, with a vision of a digital future. We conclude this volume with a review of the current developments of computer vision, which increasingly underpin our day-to-day experience of surveillance and of automation.

In the opening chapter, Professor Dan-Eric Nilsson surveys the evolution of eyes. Eyes have evolved many times in many different animal lineages. It is remarkable, however, that all eyes share similar structures in their photoreceptor cells, which all incorporate the same light-sensitive proteins (opsins). This common heritage reveals an ancient origin for vision, and we can conclude that the earliest eyes emerged sometime between 650 and 550 million years ago. Nilsson reviews the multiple strands of eye evolution, from the early animals which could only sense the presence of light, to those which could form crude images of their surroundings, and culminating in animals with an acute sense of vision which enables complex interactions with other animals. At each stage of this development it is possible to identify an adaptive advantage which continued to drive evolutionary progress towards today's diverse multitude of sophisticated eyes.

In the second chapter, Professor Paul Fletcher discusses the significance of visions and other 'illusions'. Our visual machinery has emerged from a process of Darwinian adaptation which emphasises not so much an accurate perception of the world, as a useful means to navigate and survive within it. In consequence, normal vision is not necessarily so far removed from illusion, hallucinations, and 'visions'. Indeed, there are many circumstances in which normal vision favours utility over accuracy, and mental expectation plays a key role in what we see. In extreme cases, induced for example by fatigue, fear, illness, or drugs, an entire reality can be created which conflicts with the reality perceived by others. Such conditions of psychosis offer important insights into the mechanisms of mind and serve to illuminate the processes involved in 'normal' vision.

In the third chapter, Professor Anya Hurlbert presents an extensive survey of the multiple roles of colour in vision. Colour is deeply subjective and can elicit strong, and often contradictory, responses from different viewers: J. M. W. Turner utterly transformed his seascape *Helvoetsluys*, in response to an artistic threat posed by a glorious Constable canvas exhibited on the opposite gallery wall, with a single blob of red lead, which he applied after his picture had been hung; Claude Monet, in order to realise his vision of Rouen Cathedral at different times of day, managed to transcend a strong innate tendency we all share, which modifies our perception of colour, to take account of the perceived, or assumed, ambient lighting conditions.

Colour has long been a topic of intellectual debate, not only in the visual arts, but also in philosophy, psychology, and physiology; how does colour create and convey meaning, and how does it evoke emotion and aesthetic appreciation?

In the fourth chapter, Professor Carlo Rovelli turns his attention to the role of vision in understanding the world. We have already observed that vision is honed by Darwinian adaptation, so it is not surprising that the world can be very different from what we see. In consequence, much of modern physics is considered obscure, and appears to depend entirely on a purely mathematical understanding of reality. Rovelli strongly disagrees with this view, and argues that visual images play a fundamental role in elucidating and creating scientific theory. Our minds, even when dealing with abstract and difficult notions, rely on images, metaphors, and ultimately on the mental mechanisms of vision itself, to distil understanding from what the world presents to us.

In Chapter 5 Dr Carolin Crawford reviews how we explore the cosmos. Since the earliest times humanity has been engaged in looking at the stars, using these observations for practical purposes including the management of farming, conducting religious rites, predicting the future, and gleaning an understanding of the world. Before Galileo these observations were conducted solely with the naked eye, but over the last few centuries telescopes have greatly enhanced what we can see and what we can learn about the world. We are now living through a golden age of astronomy. Most remarkably, astronomical observation has provided a window into the past, and with modern instruments we can look far back into the history of the Universe. The earliest observable structure, the Cosmic Microwave Background, dates from about 400,000 years after the Big Bang, which amounts to about 0.03 per cent of the age of the Universe, and our investigations are likely to be extended even further back in time as gravitational-wave imagery comes into its own. Crawford describes the next generation of telescopes, both ground-based and space-based, which will provide the data to further accelerate the progress made over recent decades in our understanding of the evolution of the cosmos and of fundamental physics more generally. These new instruments observe not only visible light, but also gamma-rays, X-rays, and far-infrared light, as well as gravitational waves and exotic particles.

In Chapter 6 we turn from exploring the Universe, and its past history, to articulating a vision of the future. While artificial intelligence (AI) is held by some to facilitate a utopia, others are fearful, and AI is frequently cited as presenting the most likely existential threat to humanity (eclipsing pandemic and famine). Ms Sophie Hackford is a futurist, with broad experience of AI including her own AI companies, and her role as director of WIRED magazine's consulting business. She expounds her belief that the way in which computers 'see' the world is becoming our dominant reality. Machines already view us in great detail; indeed, Hackford believes we are turning the world into a computer, and she describes how this digital 'mirrorworld' is already heavily influencing our lives. We are actively creating a physical Internet, and before long, avatars may represent us in our interactions with corporations and with society, and autonomous companies may soon become significant players in the economy.

In the final chapter, Professor Andrew Blake addresses computer vision, which is already ubiquitous, and is rapidly infiltrating much of our existence. Professor Li Fei-Fei, Google's Chief AI Scientist, recently went so far as to describe computer vision as 'AI's killer app'. Computer vision encompasses far more than mere camera surveillance; it incorporates the basis for decision-making derived from a digital analysis of images. This technology provides the means to steer an autonomous vehicle, decide on the health state of an individual, direct surgical procedures, or direct a military drone to a human target, as well as an ever-expanding list of other applications. Blake explores the reliability of these inferences, and the degree to which it is safe, or dangerous, to use the information gleaned from computer vision. Many of the vision algorithms are based on neural networks, where decisions are often opaque since the algorithmic path to a decision is difficult to discern; this can make it difficult to trust a decision. Moreover, it has been demonstrated that neural networks can be hacked by 'adversarial' counterexamples, which points to a more general fragility in a network's operation. How then, can we be sure that a computer makes good visual judgements and decisions?

1 The Evolution of Eyes

DAN-ERIC NILSSON

Eyes abound in the animal kingdom. Some are as large as basketballs and others are just fractions of a millimetre. Eyes also come in many different types, such as the compound eyes of insects, the mirror eyes of scallops, or our own camera-like eyes. Common to all animal eyes is that they serve the same fundamental role of collecting external information from light for guiding the animal's behaviour. But behaviours vary tremendously across the animal kingdom, and it turns out this is the key to understanding how eyes evolved. In this chapter we will take a tour from the first animals that could only sense the presence of light, to those that saw the first crude image of the world, and finally to animals that use acute vision for interacting with other animals. Amazingly, all these stages of eye evolution still exist in animals living today, and this is how we can unravel the evolution of behaviours that has been the driving force behind eye evolution.

The Astonishing Diversity of Eyes

The human eye is in no way special. We share its general building plan, and the way it develops in the embryo, with the other vertebrates: mammals, birds, reptiles, amphibians, and fish (Figures 1.1(a)–(d)). The striking similarities between all vertebrate eyes tell us they date back to a single common ancestor which had a typical vertebrate eye [1]. But how is it with all other eyes in the animal kingdom, such as the eyes of the octopus, or the compound eyes of insects? It turns out these eyes differ in fundamental ways from each other and even more so from the eyes of vertebrates, suggesting independent origins of vision in the animal kingdom [2–7]. Yet, the molecule responsible for sensing light, a form

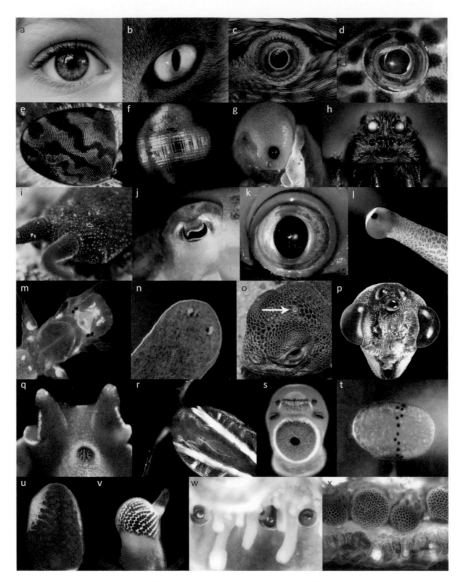

FIGURE 1.1 The diversity of animal eyes: (a)–(d), camera-type eyes of
vertebrates: human, cat, bird (parrot), and fish (coral cod); (e), insect (horse fly) compound
eye; (f), crustacean (prawn) compound eye; (g), camera-type eye in an insect (sawfly) larva;
(h), multiple pairs of camera-type eyes in a wolf spider; (i), low-resolution simple eye in a
velvet worm; (j) and (k) camera-type eyes in cuttlefish and squid; (l), low-resolution simple
eye in a snail; (m), two pairs of low-resolution simple eyes in a juvenile ragworm; (n),
lensless cup-eyes in a flatworm; (o), single parietal eye in the midline of a lizard head;

of vitamin A, seem to have been recruited for vision only once, because it is present in all animal eyes [6, 8–10]. Before we try to solve this apparent contradiction, we will first complicate the matter further by presenting a short survey of the various eye types found among invertebrate animals.

Among the most prominent eyes of any animals are those of cephalopods (octopus, squid, and cuttlefish, Figures 1.1(j) and (k)). These are animals with large eyes and excellent vision. There is a lens focusing a sharp image on a retina, just as in our own eyes. Although the shape of the lens and the eyeball differs from that of a human eye, it is practically indistinguishable from the eyes of fish or other vertebrates using vision under water. Vertebrates and cephalopods obviously share the same optical principle and functional eye design. But there are ancient differences. All parts of the cephalopod eye develop from the skin, whereas vertebrate eyes develop mainly from the brain, with only the lens and cornea originating from the skin. The retinas of cephalopod and vertebrate eyes thus have different origins. A consequence of this is that the axons sending information to the brain sensibly leave the back of the retina in cephalopod eyes but exit in the opposite direction in vertebrate eyes. This is why our retina is facing backwards, and why we need a blind spot where the axons can leave the eye.

The differences between cephalopod and vertebrate eyes are even more fundamental. The photoreceptor cells that detect the light are called rods and cones in vertebrate eyes. A close inspection of these cells reveal that they keep the visual pigment in a structure derived from a cilium. Normally, cilia are tiny motile hairs used to propel cells or the liquid around them, but in rods and cones, cilia have become immotile and strongly modified into large stacks of membrane filled with visual

---

CAPTION FOR FIGURE 1.1 (cont.) (p), group of three median eyes (dorsal ocelli) between the compound eyes of a bull ant; (q), the median compound eye in a marine flatworm; (r), directional photoreceptors in the midline of a copepod crustacean; (s), two low-resolution simple eyes and four lensless cup-eyes on a sensory club of a box jellyfish; (t), directional photoreceptors in a ring around the waist of a box jellyfish larva; (u), compound eye at the arm tip of a starfish; (v), compound eye on a tentacle of a fan worm; (w), concave mirror eyes along the mantle edge of a scallop; (x), compound eyes along the mantle edge of an ark clam.

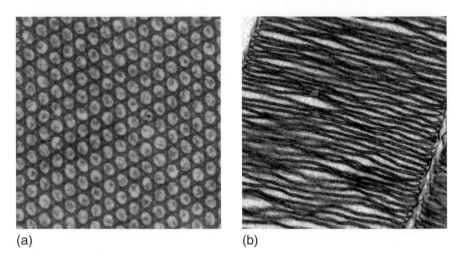

(a) (b)

FIGURE 1.2 Electron micrographs of sections through the eleborate membrane structures in an insect rhabdom (a) and a vertebrate rod photoreceptor (b). The rhabdom consists of densely packed microvilli (here cross sectioned), and the rod is a modified cilium with stacks of membrane discs.

pigment. The cephalopod retina does not have rods and cones, but photoreceptor cells of a completely different kind. Here, it is thousands of tiny membrane fingers, microvilli, that house the visual pigment in structures called rhabdoms. Microvilli without visual pigment are common in many other cells where a large surface area is needed. Consequently, photoreceptor cells of cephalopods and vertebrates have recruited two unrelated structures for elaborating the cell membrane, cilia and microvilli (Figure 1.2). Cilia use microtubules to shape the membrane, whereas microvilli use actin filaments. Thus, cilia and microvilli are fundamentally different solutions for membrane elaboration, meaning that the photoreceptor cells in vertebrate and cephalopod eyes have independently evolved the ability to pack large amounts of visual pigment in elaborate membrane structures. Rods and cones of vertebrate eyes are referred to as ciliary photoreceptors, whereas cephalopod eyes have photoreceptors referred to as rhabdomeric (rhabdoms are made of densely packed microvilli with visual pigment in the microvillar membrane).

There can be no doubt that vertebrate and cephalopod eyes have reached the same optical solution from very different origins. What about other animals? Are there more types of photoreceptor cells? It turns out that nearly all animal photoreceptors can be classified as either ciliary or rhabdomeric. The compound eyes of insects and crustaceans (Figures 1.1(e) and (f)) have rhabdomeric photoreceptors, but here the optical solution is radically different from that of cephalopod and vertebrate eyes. In compound eyes there is one lens for each pixel in the image. It is a common misconception that insects see as many images as they have facets in their eyes. That is not so. Insect vision is composed of pixels, just as human vision is. But the way light reaches the array of pixels to form an image is fundamentally different between compound eyes and simple eyes such as the camera-type eyes of vertebrates and cephalopods.

Insects and crustaceans are examples of arthropods. But not all arthropods have compound eyes. Spiders, for instance, have four pairs of camera-type eyes, and some insect larvae have a single pair of simple lens eyes (Figures 1.1(g) and (h)). Snails also have a pair of simple lens eyes, whereas rag worms have two pairs of such eyes, and planarian flatworms have one or more pairs of lens-less simple eyes (Figures 1.1(l)–(n)). The eyes mentioned so far are paired organs placed on the sides of the head, but there are also median, unpaired, eyes such as the parietal eye of lizards, the dorsal ocelli (a type of lens eyes) of insects, located between the compound eyes, or the median compound eyes present in some marine flatworms (Figures 1.1(o)–(q)).

There are eyes with even more unusual positions. Box jellyfish have no head, but well-developed lens eyes on four positions between the tentacles (Figure 1.1(s)). Starfish have compound eyes at their arm tips, and fan worms have compound eyes on their feeding tentacles (Figures 1.1(u) and (v)). Scallops have hundreds of eyes with mirror optics looking out along the shell opening (Figure 1.1(w)). Other clams have a row of lens-less compound eyes in the same position (Figure 1.1(x)). Chitons have simple eyes sprinkled all over their back, and sea urchins are covered in dispersed photoreceptors that make the entire body act as a large compound eye.

These examples give an idea of the enormous diversity of eyes in animals. Eyes can be simple or compound, have lenses or no lenses, use rhabdomeric or ciliary photoreceptors, be paired or median, and placed on the head or elsewhere [11–13]. It is this diversity we have to account for when trying to reconstruct how eyes evolved. Initially it may seem to complicate the matter, but this diversity in fact holds the key to the origin of eye evolution.

How Did It All Start?

A single question will lead us on the right track: is there any cell type present in all eyes, from the simplest to the most advanced, that could be useful on its own, without all the other parts of the eyes? The answer is yes, the *photoreceptor cells* could obviously serve important functions on their own. Without the presence of any ocular structures, they could tell when it is day and when it is night, which is crucial for selecting the best time for different behaviours and for setting the biological clock. For primitive animals in water, photoreceptor cells are also essential for sensing their depth in the sea, in order to be where food is most plentiful, and for avoiding excessive levels of harmful ultraviolet light close to the surface.

Most animals still have such a sense, which just measures the levels of ambient light, and often use it to control the levels of sleep hormone, melatonin. Photoreceptor cells with this function are often located in the brain. We have such cells in our eye – possibly because it is too dark in our brain – but they are not the rods and cones we use for vision. This sense is called *non-directional light sensitivity* because it is based on measuring ambient light levels irrespective of the directions light comes from [5, 7, 14].

The first animals would have been very small creatures that either crawled or swam with cilia. They are likely to have had a chemical sense and a tactile sense to orient them towards suitable places in the environment. If they had any light sensitivity it must have started as non-directional sensing of light. Even though this was a very long time ago, evolution has left clues that now allow us to reconstruct these first steps on the way to eyes and vision. By far the most important clue is the mechanism by which light is detected. In all animal eyes this is done by a very special type of receptor proteins located in the cell membrane. These receptor proteins are called opsins [8–10], where rhodopsin is the version found in the rods of our own retina.

Opsins in turn, are part of a huge family of signalling proteins called G-protein-coupled receptor proteins (GPCRs) that exist in animals, fungi, and protists but not in bacteria. Their general function is to detect specific chemicals and forward the information to the rest of the cell [6, 15]. There are thousands of different types of GPCRs that specifically bind to and detect a huge range of different chemicals. Opsins bind to derivatives of vitamin A, and it is the vitamin A molecule that is sensitive to light. A bent isoform of Vitamin A, called 11-cis-retinal, can absorb the energy in a single particle of light, a photon, which flips the molecule into its straight isoform, all-trans-retinal. Each opsin molecule can hold a single vitamin A molecule in a pocket and signals to the cell when the vitamin A changes from its bent to its straight isoform. This means we actually sense light by chemoreception, or, in other words, our eyes detect light by tasting a photoproduct of vitamin A.

Fundamental similarities of all animal opsins reveal that they originated once in the very early phases of animal evolution. Just to confuse the matter, there are other opsins around. Some fungi have been reported to have opsins, but these are unrelated to animal opsins. Bacteria also have molecules called opsins, but these do not belong to the GPCRs. Thus, the unique family of animal opsins tells us that this mode of light sensitivity arose very early in animal evolution [16].

The closest GPCR relative to animal opsins is the melatonin receptor, i.e. a GPCR that detects the presence of the sleep hormone melatonin [17, 18]. This is interesting because melatonin levels are controlled by light detection with opsins. We know that melatonin receptors predate the first animal because they are found also in protists. It is possible and even likely that melatonin receptors served as light receptors before opsins evolved [7]. The reason is that melatonin is itself light sensitive and is destroyed by oxidation when illuminated. With a constant synthesis of melatonin, the concentrations would thus be high at night and low during the day. In such a simple system the melatonin receptor would have served as an ancient light receptor predating the opsins. But melatonin-based light receptors suffer from several limitations. Because melatonin is destroyed by light, new melatonin would constantly have to be synthesised, and changes in the light intensity during the day would be hard to detect because of low melatonin levels. Melatonin is also primarily sensitive to ultraviolet light, which is never as bright as the longer, visible, wavelengths.

The fact that melatonin receptors are the closest relatives to opsins suggests that it was a modification of an ancient melatonin receptor that gave rise to the first opsin [7]. By changing from binding melatonin to binding vitamin A aldehyde, the receptor became a far more efficient light receptor. Whereas melatonin is destroyed by light, vitamin A aldehyde just flips from bent to straight, and this can easily be flipped back such that the molecule can be recycled indefinitely. This is not only more economical, but also allows automatic changes of sensitivity such that small differences of intensity can be discriminated under both dim and bright conditions. Modifications of the binding pocket inside the opsin protein also allow changes of the wavelength sensitivity over a large range from ultraviolet to red. There are thus numerous benefits of replacing melatonin with vitamin A, and the first opsin would have been a major improvement in light detection.

A close look at the phylogenetic relationship of melatonin receptors and opsins among the major groups of animals reveals some interesting facts (Figure 1.3). Very shortly after the first animal opsin evolved there was a major radiation into different families of opsins, supporting the idea that the first opsin opened up new possibilities, which were rapidly exploited. Sponges were the first animal group to split from the rest of the animals, and sponges do not have any opsins – but they do have melatonin receptors (Figure 1.3). This implies that the first opsin evolved in animals after the split between sponges and other animals but before other animals diverged in the present major groups. Being able to pinpoint opsin origin that precisely also allows a reliable timing of this event. By comparisons of animal evolution, which can be reliably timed on the basis of the entire genome [19], we can work out when the first opsin evolved (Figure 1.4). This turns out to have been about 800 million years ago, and it was the starting point of an evolutionary journey towards eyes and vision.

What Happened after the First Opsin Evolved?

Melatonin receptors would have been adequate for telling day from night, suggesting that the rapid radiation of opsins shortly after their introduction enabled new functions that had not been possible with melatonin receptors. What might these new functions have been? There is only so

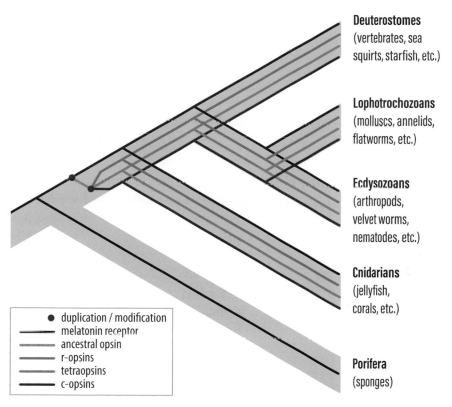

Deuterostomes
(vertebrates, sea
squirts, starfish, etc.)

Lophotrochozoans
(molluscs, annelids,
flatworms, etc.)

Ecdysozoans
(arthropods,
velvet worms,
nematodes, etc.)

Cnidarians
(jellyfish,
corals, etc.)

Porifera
(sponges)

- duplication / modification
— melatonin receptor
— ancestral opsin
— r-opsins
— tetraopsins
— c-opsins

FIGURE 1.3 The divergence of opsins in major animal groups.

much a non-directional photoreceptor can be used for. Reading the general ambient intensity could inform about the time of day, the water depth, or possibly a sudden shadow. A photoreceptor cell could be used for entirely different tasks if it provided information about how light is distributed in the environment. Such information would allow movement towards bright or dark parts of the environment (phototaxis), and greatly enhance an animal's ability to find optimal positions in the environment [5, 7]. Although non-directional photoreceptors are unable to provide guidance for phototaxis, it only takes some dark shielding pigment to make a photoreceptor cell directional (Figure 1.5(b)).

The combination of a photoreceptor cell and a dark pigment cell is a common feature in many of the small, wormlike animals that make up

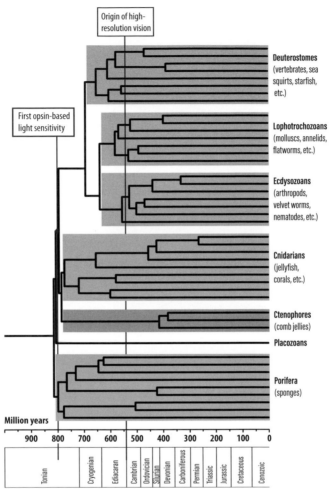

FIGURE 1.4 A timed phylogeny of major animal groups. Geological periods are indicated below the time scale.

much of the animal kingdom (see an example in Figure 1.6). In bilaterally symmetric animals, such combinations typically occur as paired organs in the frontal part of the animal. By turning the head or body, directional photoreceptors are made to scan the environment and can guide the animal towards brighter or dimmer parts of the environment. Directional photoreceptors can also be used to determine

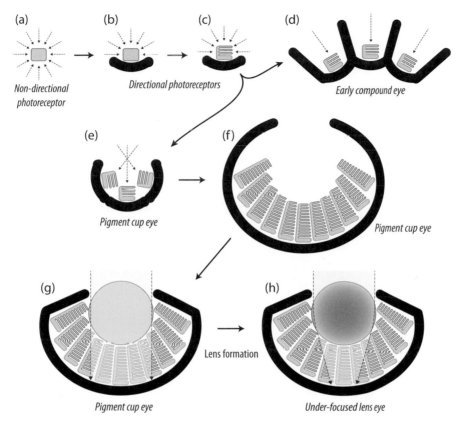

FIGURE 1.5 Schematic drawings of major stages in the evolution of photoreception, starting with non-directional photoreception (a), directional photoreception without membrane stacking (b), the same with membrane stacking (c), a low-resolution compound eye (d), a low-resolution cup eye (single-chambered eye) (e), a larger version of a low-resolution cup eye (f), a cup eye with a protective vitreous mass filling the cavity above the photoreceptors (g), and a more elaborate low-resolution eye where the vitreous mass had turned into a lens to produce an under-focused lens eye (h).

what is up and down for control of body posture. Such optical statocysts are known from many invertebrates, often as unpaired median organs (e.g. Figures 1.1(p)–(r)). Finding optimal locations and body orientations in the environment would have been major competitive advantages offered by the introduction of directional photoreceptors.

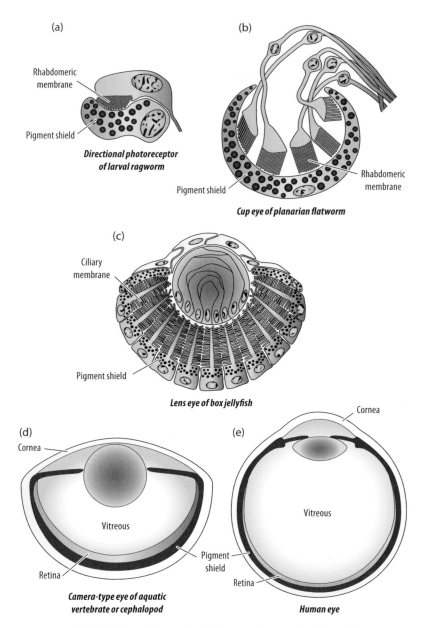

(a)

Rhabdomeric membrane

Pigment shield

Directional photoreceptor of larval ragworm

(b)

Pigment shield

Rhabdomeric membrane

Cup eye of planarian flatworm

(c)

Ciliary membrane

Pigment shield

Lens eye of box jellyfish

(d)

Cornea

Vitreous

Retina

Pigment shield

Camera-type eye of aquatic vertebrate or cephalopod

(e)

Cornea

Vitreous

Retina

Human eye

FIGURE 1.6 Examples of the morphology of different stages in eye evolution. Colours indicate photoreceptor cells or retina (green), pigment screen (brown, black granules), optics (blue), and external protective tissue (yellow). The directional photoreceptor (a) and the low-resolution eyes (b) and (c) are all less than 0.1 mm in diameter, whereas typical high-resolution eyes, (d) and (e), are 10–1,000 times larger (1 mm–10 cm).

To change a non-directional photoreceptor into a directional one only takes some screening pigment, but, in order to optimise the directional photoreceptor to its new tasks, a few additional modifications will be required. The typical task of non-directional photoreceptors is to detect the daily light intensity slowly changing over a huge range of some 100 million times (eight orders of magnitude) from bright sunlight to a moonless night. Scanning a directional photoreceptor cell to guide phototaxis is a very different task where the speed must be very much faster [20], and the range of intensities to be discriminated varies over a much smaller range of two orders of magnitude from the dimmest to the brightest direction within a scene. But the position of this restricted range slides up and down the daily variation of eight orders of magnitude. This means the photoreceptor would have to speed up its response massively and also introduce some mechanism of shifting its gain to cope with the large variations of light intensity at dusk and dawn, and under different weather conditions [21].

Basically, the photoreceptor cell would have to change from detecting slow changes of absolute intensity to discriminating much faster changes of relative intensity within a scene. For a photoreceptor cell informing about the time of day or depth in the sea, adaptation would be detrimental, but, for comparing intensities at different directions in the environment, adaptation is a necessity. The distinction between adapting and non-adapting receptor cells is not unique to light receptors: most tactile receptors would need an adaptation mechanism, although adaptation would defeat the purpose of a receptor for noxious heat. The types of molecular machinery behind these two modes of operation are likely to have originated in other senses before they appeared in photoreceptors. An important conclusion from this is that the same photoreceptor cell cannot inform about the time of day and also guide phototaxis. There is thus a need for different types of photoreceptors to support different parts of more complex behaviours, and this is a likely reason for the early radiation of opsin types (Figure 1.3). In the next section we shall discover another principle that had a major influence on the evolution of photoreceptor cells.

A Problem of Photon Shortage Guides Evolution

Photoreceptor cells can detect light only by absorbing it, and then light behaves as particles, called photons. A consequence of this is that the detected brightness is in principle based on the number of photons contributing to the neural signal in photoreceptor cells. Even at a constant light intensity, the photons arrive randomly, very much like drops of rain, and repeated attempts to determine the intensity will result in slightly varying photon counts. More photons in each count will result in a more precise knowledge about how bright it is. The phenomenon is known as photon noise, and it sets a limit to the smallest differences in light intensity that can be discriminated. As an example, if the sensory task is to signal the arrival of dawn, the photoreceptor cell must be able to detect an increase in light intensity that exceeds the photon noise. The slowness of intensity changes means that photoreceptor cells can respond very slowly, effectively counting photons over 10 minutes or more. Then it can be calculated that a small photoreceptor cell with opsin densely packed in its membrane will count enough photons to just discriminate the intensity change from starlight to the first light of dawn [7].

   This reasoning can now be used as a tool to understand eye evolution. If we calculate the ambient intensity that is needed for a directional photoreceptor to guide an animal to a dark shelter (negative phototaxis), we will find that the conditions are very different from the previous case of spotting the first light of dawn. Intensities at different directions within an environment at any given time differ much less than the huge change between day and night. Consequently, much smaller differences in intensity must be discriminated, and this requires more photons in each of the samples that are to be compared. But, to guide movement, the cell must be very much faster, with an integration time in the range of a second (almost three orders of magnitude faster than the change at dusk and dawn). On top of this, the directionality of the photoreceptor allows it to pick up light only from within a restricted angle. If we do the numbers, it turns out that the phototaxis task would work only down to mid dusk on a sunny day on land [7]. At a few metres depth in coastal water it would barely work on an overcast day, and at 10 metres depth it would not work at all. There are simply not enough photons in natural habitats to allow discrimination of light in different directions, using the same photoreceptor that we know can signal the onset of dawn.

To cure the directional photoreceptor of its shortage of photons, light collection must become more efficient. The main problem is that no more than 0.02 per cent of the light will be absorbed on passing through a membrane packed with opsin molecules. The remaining light, 99.8 per cent, will be lost by the photoreceptor cell. Clearly, this provides an opportunity for major improvements by simply folding the membrane into stacks with numerous layers. This is exactly what we find in most directional photoreceptors. Rather than spreading the opsin over a non-specialised membrane, these photoreceptor cells confine their opsin to heavily elaborated membrane structures. With the possibility of thousands of membrane layers, a photoreceptor cell can improve its sensitivity by absorbing much of the light reaching it. These membrane elaborations are evolutionary responses to the shortage of photons caused by the introduction of directional photoreception.

Ancient Alternative Solutions

As we recall, there are two major ways to elaborate the cell membrane, cilia and microvilli, and these existed well before any photoreceptor cells evolved in animals. The distinctions between cilia and microvilli are both functional and structural. Cilia are originally motile structures supported by microtubules, whereas microvilli are cell surface extensions supported by actin filaments. Since they are based on two different types of cyto-skeleton, cilia and microvilli are distinct structures with no intermediates. Because photoreceptor cells with membrane elaborations are of two basic types, using either cilia or microvilli to hold the opsin, we can conclude that directional photoreception must have evolved independently at least twice, generating ciliary and rhabdomeric (microvillar) photoreceptors (Figure 1.2).

This conclusion is further supported by the fact that specific families of opsin proteins are associated with cilia and microvilli, respectively. There is thus a distinct class of ciliary opsins (c-opsins) found in ciliary photoreceptors and an equally distinct class of rhabdomeric opsins (r-opsins) found in rhabdomeric photoreceptors. Also, the molecular transduction mechanisms responsible for sensitivity adjustment (light/dark-adaptation), which became necessary in the first directional photoreceptors, are of fundamentally different types in ciliary and rhabdomeric photoreceptor cells [6].

When these photoreceptor types were first described by electron microscopy, it was thought that vertebrates had ciliary photoreceptors and invertebrates had the rhabdomeric type [22]. But exceptions were soon discovered, and it eventually turned out that all major animal groups possess both types of photoreceptor cells [2–4, 23], except sponges, which have no opsin-based photoreceptors at all [24]. It follows that both types of photoreceptors, ciliary and rhabdomeric, evolved independently as directional photoreceptors in a common animal ancestor after sponges had split off (see Figure 1.3). From a timed phylogenetic tree of animals (Figure 1.4), we can pinpoint the life of this last common ancestor to shortly after 800 million years ago.

The Origin of Eyes

With both non-directional and directional photoreceptors (Figures 1.5(a)–(c)), the early animal ancestor could use light to control a range of simple behaviours such as regulating activity according to the daily light cycle, regulating the animal's depth in water, moving towards or away from light, or orienting the body to the up–down gradient of light intensity. Because each new function required specific performance properties of the photoreceptor cells, this evolution led to several distinct types of photoreceptor cells, each specialised for a particular task. In the more simply organised animals that remain today, such as jellyfish, each behaviour that is controlled by light appears to have its own group of photoreceptor cells in its own specific place [11, 21, 25]. Only later was light detection centralised into specific organs – eyes.

So how did eyes evolve from the photoreceptors we have discussed so far? The step is actually rather short. Directional photoreceptors operate by guiding the animal towards dimmer or brighter directions to find the optimal position in the environment. The animal can do this only by continuously turning its body until it aims in a direction with a desired light intensity. To find the overall light gradients in the environment, the photoreceptor must pick up light from a very large angle, roughly a hemisphere, which also leads to limited accuracy in finding the best direction. Improving this situation requires narrower angles of sensitivity, which in turn requires much more scanning to cover all angles. This problem is easily solved by multiplying the directional photoreceptor and having the photoreceptors point in different directions (Figure 1.5(d)). This also abolishes the need for the scanning body movements since information about the intensity in different directions can be obtained by comparing the signals from the different photoreceptors. Such spatial resolution within a group of photoreceptors is the basic definition of an eye.

If we use terminology from digital cameras, each directional photoreceptor with a unique field of view corresponds to one image pixel. It is obvious that a small number of directional photoreceptors will provide only a very crude image of the environment, but it is more accurate than the single-directional photoreceptor, scanning is no longer necessary, and, most importantly, resolution can be improved by just continuing to multiply the number of photoreceptors and correspondingly narrow their angle of sensitivity. Complicating factors are that a new type of neural mechanism comparing neighbouring photoreceptors must be introduced, and the behavioural control must change from scanning to steering. This calls for centralised neural control – a brain. Elaborations of the developmental genetics controlling cell position and orientation must, of course, also be introduced at this point. A beautiful recapitulation of the transition from directional photoreception to spatial vision, and the first formation of a brain, can be seen in larval stages of modern ragworms [20].

Two Solutions to Imaging

Multiplying directional photoreceptors to make a low-resolution image immediately forces a choice between two alternatives. The photoreceptors can form a common cup or point away from one another (Figures 1.5(d) and (e)). The common cup alternative will lead to a single-chambered eye, and the other alternative will lead to a compound eye. We know from the occurrence of eyes in different animal groups today that both directions were initiated numerous times independently. The single-chambered eyes of jellyfish, molluscs, and vertebrates originated independently from directional photoreceptors, as can be judged from the use of different types of opsins, different transduction mechanisms, development from different tissues, and the use of different principles for membrane stacking. The same is true for compound eyes in groups such as insects, fan-worms, and starfish, which also show convincing evidence of having independently evolved from different types of directional photoreceptors [21]. But, before we continue to the sophisticated eyes of vertebrates, cephalopods, and insects, we shall ask some important questions about the low-resolution eyes that evolved from directional photoreceptors.

When Did the First Eye See the World?

Because eyes obviously evolved independently many times from directional photoreceptors, it is not easy to work out which animal group was the first to take this step. It could even be that the first animals with eyes belong to a group that became extinct long time ago and left no traces. But if we assume that the first eye evolved within an animal group that still exists today and that their present eyes date back to the first eye, then the phylogenetic tree of Figure 1.4 can give us some indication. Because the eyes in different animal groups differ in terms of being compound or single-chambered, using rhabdomeric or ciliary photoreceptors, etc., we can conclude that eyes must have evolved after the last common ancestor of groups that independently acquired eyes, but before the split of groups that share the same type of eye. Using this kind of reasoning, the first eyes probably arose sometime between 650 and 550 million years ago. Then there are also many groups where eyes evolved much later.

What Are Low-Resolution Eyes Used For?

In every group where eyes evolved, the first eyes must have started with just a small number of photoreceptor cells (Figures 1.5(d) and (e) and 1.6(b)), and consequently been able to resolve an image with only a small number of pixels. For finding a suitably lit environment, this is still better than just a non-directional photoreceptor. But, as soon as the multiplication of units was initiated, it would be developmentally simple to keep adding new units to bring up the number of photoreceptor cells and thus the number of image pixels. It would seem that once this process was initiated, high-resolution vision would evolve rapidly. But in most branches of the animal kingdom that possess eyes today (about half of the roughly 30 animal phyla), all but three still have only low-resolution eyes with anything between 10 and a couple of hundred photoreceptor cells. These eyes are tiny, less than a millimetre across, and not very conspicuous.

Clearly, low-resolution vision must be very useful and there must be reasons not to continue eye evolution beyond a certain level of acuity. Because examples of these *low-resolution eyes* are so abundant, their roles have now been studied in several animal groups such as jellyfish, flatworms, velvet worms, and starfish, and the outcome is simple and straightforward [21, 26–30]. Low-resolution eyes are used for finding a suitable habitat and for positioning of the animal within that habitat. For these tasks, it is sufficient to see large stationary structures in the environment. Low-resolution vision, i.e. a few hundred pixels and less, is adequate for moving about, avoiding collisions with objects, finding a free path, moving in and out of shelters, finding areas where food is likely to be found by other senses, or keeping a straight course (an excellent strategy to get out of an unfavourable environment). Resolution (acuity) of a few hundred pixels or less is not good enough to see other animals at any distance beyond immediate reach. Low-resolution eyes therefore cannot be used to see prey, predators, or individuals of the same species.

What animals with low-resolution vision are able to see appears poor indeed to us, but is perfectly adequate to support the visually guided behaviours of these animals. In Figure 1.7, vision in a velvet worm is shown as an example. These worms are found in the southern hemisphere. They belong to the phylum Onychophora, which is a sister group of the arthropods. Velvet worms are active at night and hunt for insects using touch and smell. They cannot see their insect prey at distances beyond immediate reach. At dawn they have to find a moist shelter under a fallen log, or else they will dry out and perish before the day is over.

FIGURE 1.7 High-resolution human vision compared with low-resolution vision in a velvet worm. Colour vision is common among the high-resolution eyes of vertebrates and insects, whereas low-resolution eyes typically are colour-blind. The low-resolution vision of the velvet worm is good enough for finding fallen logs or other shelters (simulation by Mikael Ljungholm).

Their tiny eyes will guide them to suitable shelters at dawn, and also guide them back out to suitable hunting grounds at dusk [30].

The Problem of Photon Shortage Returns: The Evolution of Lenses

In single-chambered eyes, resolution can be improved by adding photo-receptors, making the eye cup deeper, and making the aperture smaller (as illustrated in Figures 1.5(e) and (f)). In compound eyes the number of units (ommatidia) must grow, and the photoreceptors must be set in correspondingly deeper pigment tubes. In both cases, the visual angle of each photoreceptor must shrink as the number of photoreceptors grows. When the animal moves, this leads to motion blur unless the integration time is also made shorter. On top of this, lower-contrast image details become of interest as resolution increases.

We recall that the transition from non-directional to directional photo-reception was associated with a sensitivity problem that was solved by creating stacks of membrane to catch a larger fraction of the light that reaches a photoreceptor cell. Now, as the resolution of the first eyes was increasing, more photons needed to be detected by each photoreceptor during each integration time (to see lower contrasts), but the narrower angles and shorter integration times have the opposite effect. The consequence of this is that improvement of resolution by adding more units requires an increasingly brighter world for the eye to function.

To some extent the need for better sensitivity can be met by deeper stacks of photoreceptor membrane, such that these structures turn into long rods. But this is not enough to bring resolution up to more than 10–50 pixels, and photon shortage would make vision impossible at dusk or at more than a few metres depth in coastal water. Better resolution and better sensitivity require a new invention. The answer is to concentrate light by focusing, or in other words to make a lens in front of the photoreceptors [5, 7]. In single-chambered eyes there is obvious space for a lens in the eye cup, and this may anyway be filled by vitreous cells protecting the eye from damaging ultraviolet light or providing structural support (Figures 1.5(g) and 1.6(c)). If these cells produce large amounts of protein, they will form a lens that concentrates light on all photoreceptors in the eye and cures the photon shortage. The lens need not be perfectly focused on the layer of photoreceptors (the retina; see Figure 1.5(h)). For low-resolution vision partial focusing will suffice. In compound eyes, increasing resolution leads to large numbers of extremely long pigment tubes that have to be perfectly straight. Here too lenses offer an excellent solution, which allows narrow visual angles in much shorter ommatidia.

Without lenses, eyes would not be able to evolve beyond the first extremely low-resolution versions still found in flatworms today (Figures 1.1(n) and 1.6(b)). Most other low-resolution eyes have some sort of lenses (Figures 1.5(h) and 1.6(c)). It is important to note that the primary purpose of the lenses in low-resolution eyes is not to make sharp images, but to bring sensitivity up to levels needed for vision at the naturally occurring light intensities on planet Earth.

From Low to High Resolution and the Birth of a
New Ecological System

Increasing the resolution such that each pixel covers only a few degrees means that other animals can be spotted at distances larger than a few body lengths. When this threshold was reached for the first time, a new ecological era was initiated. With an instantaneous long-range sense, predation becomes a lucrative business. And it is easily made even more effective by continuing to improve resolution. But here, the problem of photon shortage strikes for the third and last time. When membrane stacking has been fully exploited and lenses have been introduced, there is only one way left to increase sensitivity beyond that of low-resolution vision. That way is simply to make the eye larger (Figures 1.6(d) and (e)). A larger lens has a larger area picking up light, and this will compensate for the smaller pixel angles in high-resolution vision. The great thing here is that this strategy has no upper limit. If the eye is large enough, any kind of resolution can be obtained at any light intensity. But the growth in eye size from the sub-millimetre low-resolution eyes must be substantial to enable long-range spotting of other animals or seeing at very low intensities. Accordingly, high-resolution eyes measure from a few millimetres to more than 25 cm in present-day animals. The largest eyes, with a diameter of 27 cm, belong to deep-sea giant squid inhabiting depths of 1,000 metres in the ocean [31].

The required growth in eye size may have been the main obstacle for adopting a visually guided predatory lifestyle. The lack of obvious animal fossils older than 540 million years suggests that earlier animals were small and soft-bodied like many of the present-day animals that have only low-resolution vision. But growing large eyes to get resolution high enough for seeing other animals is not entirely straightforward. High resolution itself means that the nervous system will have to process information from a very large number of photoreceptor cells. And, to see other animals, there has to be a sizeable brain with new types of neural circuits that can detect small moving objects against a cluttered background [32]. For this information to be useful, the animal will also have to evolve means of efficient locomotion with muscles and a skeleton to attach them to, and of course motor circuits in the brain that can

orchestrate adequate behaviours in response to the new visual information. All of this together means that a large, mobile animal with complex
behaviours would have to evolve from a small, slow-moving animal with
much simpler behaviours. In terms of evolutionary change, this is quite
a revolution.

For the species that underwent this change, it would have implied an
entirely new and very active lifestyle. For the ecological system, the first
visually guided predator would have introduced completely new selective
pressures on other species that tried to avoid ending up on the dinner
plate. Potential evolutionary responses would have been to develop equally
good vision and efficient mobility to escape the predators, starting an arms
race with vision and mobility. Other alternatives would be to develop
protective armour, or change to a burrowing lifestyle to avoid the visually
guided predators. All these evolutionary innovations would have left
obvious traces in the fossil record. And they did indeed. The event called
the Cambrian explosion, about 540 million years ago, displays everything
expected from the introduction of visually guided predation: the first
macroscopic animals with prominent eyes, fins, and walking appendages,
internal as well as external skeletons for muscle attachment, protective
shells, body armour, and spikes, and the first deep-burrowing animals. All
of these appear suddenly in the fossil record from the early Cambrian era,
suggesting that it really was a major turnover forcing the entire ecological
community to undergo a dramatic change (2, 33, 34). All just because some
animals went from low-resolution vision to high-resolution vision.

Most of the early Cambrian fossil animals appear to have had an
exoskeleton and compound eyes, but that may be a false bias because
they are more prone to form fossils than other types of animals and
types of eyes. Even the vertebrate eye probably arose in the early
Cambrian, whereas cephalopods may have acquired their high-
resolution eyes a little later. The rest of the animal kingdom kept their
low-resolution eyes, and still have them today. Many of these animals
are tiny and inconspicuous, which of course helps in a world full of
visually guided predators. We still live in the ecological system that
started 540 million years ago. In principle, not much novelty has been
introduced in vision for the past 500 million years, even though animals
have in that time evolved into many new forms and conquered the land.

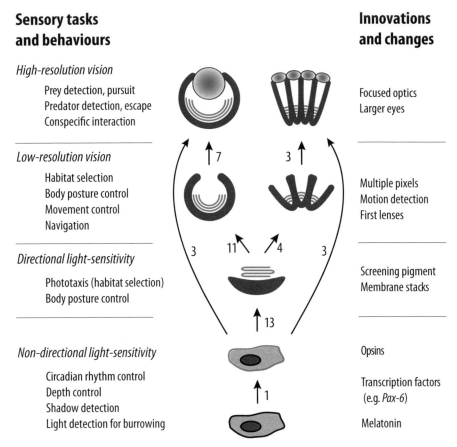

Sensory tasks and behaviours

High-resolution vision

Prey detection, pursuit
Predator detection, escape
Conspecific interaction

Low-resolution vision

Habitat selection
Body posture control
Movement control
Navigation

Directional light-sensitivity

Phototaxis (habitat selection)
Body posture control

Non-directional light-sensitivity

Circadian rhythm control
Depth control
Shadow detection
Light detection for burrowing

Innovations and changes

Focused optics
Larger eyes

Multiple pixels
Motion detection
First lenses

Screening pigment
Membrane stacks

Opsins

Transcription factors
(e.g. *Pax-6*)

Melatonin

FIGURE 1.8 A schematic diagram of the four key stages in eye evolution. The stages of evolution of new behaviours are listed to the left, and associated innovations and changes are listed to the right.

Some later cases of new eye evolution have occurred in animals such as clams and fan worms (Figures 1.1(v)–(x)), where the original eyes were lost and non-directional shadow detectors were recruited to form new eyes for predator detection [35–37].

To sum up this long story, eye evolution followed a series of four steps: non-directional light sensitivity, directional light sensitivity, low-resolution vision, and finally high-resolution vision. In three specific phases, photon shortage intervened in the evolutionary process and led to membrane stacking, lenses, and large eyes. The evolutionary process leading to eyes started when the first opsin evolved about 800 million years ago and reached the final stage of high-resolution vision some 540 million years ago. The first opsin protein evolved in a common ancestor after sponges had split off the animal family tree. Except for the origin of opsins, all the major steps of eye evolution have occurred multiple times, and some rough counts are presented in Figure 1.8. It was the competition for efficient behaviours that drove evolution from simple light sensing to acute vision. Photoreception and vision have been pivotal for the evolution of animals, and even our own existence is a consequence of eye evolution.

The discovery of identical control genes for eye development (*Pax-6*) in vertebrates and insects [38] was originally taken as evidence for a single origin of all animal eyes, but overwhelming evidence to the contrary [2–7, 12, 16, 24, 25, 36, 39] clearly demonstrates that each of the transitions between the four key stages has occurred independently numerous times (a conservative count of the number of independent transitions is indicated next to the arrows in Figure 1.8). The need for opsin expression in specific cells is the obvious reason for conserved developmental genes.

References

[1] Lamb, T. D., Pugh Jr, E. N., and Collin, S. P. The origin of the vertebrate eye. *Evol. Edu. Outreach* 2008; 1: 415–426. DOI https://doi.org/10.1007/s12052–008-0091-2

[2] Nilsson, D.-E. Eye ancestry: Old genes for new eyes. *Curr. Biol.* 1996; 6: 39–42. DOI https://doi.org/10.1016/S0960–9822(02)00417-7

[3] Nilsson, D.-E. Eye evolution: A question of genetic promiscuity. *Curr. Opin. Neurobiol.* 2004; 14: 407–414. DOI https://doi.org/10.1016/j.cub.2005.01.027

[4] Nilsson, D.-E. Photoreceptor evolution: Ancient siblings serve different tasks. *Curr. Biol.* 2005; 15: R94–R96. DOI https://doi.org/10.1016/j.cub.2005.01.027

[5] Nilsson, D.-E. The evolution of eyes and visually guided behaviour. *Phil. Trans. R. Soc. B* 2009; 364: 2833–2847. DOI https://doi.org/10.1098/rstb.2009.0083

[6] Fain, G. L., Hardie, R., and Laughlin, S. B. Phototransduction and the evolution of photoreceptors. *Curr. Biol.* 2010; 20: R114–R124. DOI https://doi.org/10.1016/j.cub.2009.12.006

[7] Nilsson, D.-E. Eye evolution and its functional basis. *Visual Neurosci.* 2013; 30: 5–20. DOI https://doi.org/10.1017/S0952523813000035

[8] Porter, M. L., Blasic, J. R., Bok, M. J., Cameron, E. G., Pringle, T. et al. Shedding new light on opsin evolution. *Phil. Trans. R. Soc. B* 2012; 279: 3–14. DOI https://doi.org/10.1098/rspb.2011.1819

[9] Liegertova, M., Pergner, J., Kozmikova, I., Fabian, P., Pombinho, A. R. et al. Cubozoan genome illuminates functional diversification of opsins and photoreceptor evolution. *Sci. Rep.* 2015; 5: 11885. DOI https://doi.org/10.1038/srep11885

[10] Ramirez, M. D., Pairett, A. N., Pankey, M. S., Serb, J. M., Speiser, D. I. et al. The last common ancestor of most bilaterian animals possessed at least nine opsins. *Genome Biol. Evol.* 2016; 8: 3640–3652. DOI https://doi.org/10.1093/gbe/evw248

[11] Land, M. F., and Nilsson, D.-E. General purpose and special purpose visual systems. In Warrant, E. J., and Nilsson, D.-E, eds. *Invertebrate Vision.* Cambridge: Cambridge University Press, 2006; 167–210.

[12] Land, M. F., and Nilsson, D.-E. *Animal Eyes*, 2nd ed. Oxford: Oxford University Press, 2012.

[13] Cronin, T. W., Johnsen, S., Marshall, N. J., and Warrant, E. J. *Visual Ecology.* Princeton, NJ and Oxford: Princeton University Press, 2014.

[14] Foster, R. G., and Hankins, M. W. Non-rod, non-cone photoreception in the vertebrates. *Prog. Retin. Eye Res.* 2002; 21: 507–527. DOI https://doi.org/10.1016/j.conb.2005.06.011

[15] de Mendoza, A., Sebe-Pedros, A., and Ruiz-Trillo, I. The evolution of the GPCR signaling system in eukaryotes: Modularity, conservation, and the transition to metazoan multicellularity. *Genome Biol. Evol.* 2014; 6: 606–619. DOI https://doi.org/10.1093/gbe/evu038

[16] Colley, N, J., and Nilsson, D.-E. Photoreception in phytoplankton. *Integ. Comp. Biol.* 2016; 56: 764–775. DOI https://doi.org/10.1093/icb/icw037

[17] D'Aniello, S., Delroisse, J., Valero-Gracia, A., Lowe, E. K., Byrne, M., et al. Opsin evolution in the Ambulacraria. *Mar. Genom.* 2015; 24: 177–183. DOI https://doi.org/10.1016/j.margen.2015.10.001

[18] Tosches, M. A., Bucher, D., Vopalensky, P., and Arendt, D. Melatonin signaling controls circadian swimming behavior in marine zooplankton. *Cell* 2014; 159: 46–57. DOI https://doi.org/10.1016/j.cell.2014.07.042

[19] Dohrmann, M., and Wörheide, G. Dating early animal evolution using phylogenomic data. *Sci. Rep.* 2017; 7: 3599. DOI https://doi .org/10.1038/s41598–017–03791–w

[20] Randel, N., and Jekely, G. Phototaxis and the origin of visual eyes. *Phil. Trans. R. Soc. B* 2016; 371: 20150042. DOI https://doi.org/10 .1098/rstb.2015.0042

[21] Nilsson, D.-E., and Bok, M. J. Low-resolution vision – at the hub of eye evolution. *Integ. Comp. Biol.* 2017; 57: 1066–1070. DOI https:// doi.org/10.1093/icb/icx120

[22] Eakin, R. M. Structure in invertebrate photoreceptors. In Autrum, H., ed. *Handbook of Sensory Physiology*, **vol. 7**. Berlin: Springer, 1972; 625–684.

[23] Arendt, D., Tessmar-Raible, K., Snyman, H., Dorresteijn, A. W., and Wittbrodt, J. Ciliary photoreceptors with a vertebrate-type opsin in an invertebrate brain. *Science* 2004; 306: 869–871. DOI https://doi.org/10.1126/science.1099955

[24] Vopalensky, P., and Kozmik, Z. Eye evolution: Common use and independent recruitment of genetic components. *Phil. Trans. R. Soc. B* 2009; 364: 2819–2832. DOI https://doi.org/10.1098/rstb.2009 .0079

[25] Picciani, N., Kerlin, J. R., Sierra, N., Swafford, A. J., Ramirez, M. D. et al. Prolific origination of eyes in Cnidaria with co-option of non-visual opsins. *Curr. Biol.* 2018; 28: 2413–2419. DOI https://doi .org/10.1016/j.cub.2018.05.055

[26] Nilsson, D.-E., Gislén, L., Coates, M. M., Skogh, C., and Garm, A. Advanced optics in a jellyfish eye. *Nature* 2005; 435: 201–205. DOI https://doi.org/10.1038/nature03484

[27] Garm, A., O'Connor, M., Parkefelt, L., and Nilsson, D.-E. Visually guided obstacle avoidance in the box jellyfish *Tripedalia cystophora and Chiropsella bronzie*. *J. Exp. Biol.* 2007; 210: 3616–3623. DOI https://doi.org/10.1242/jeb.004044

[28] Garm, A., Oskarsson, M., and Nilsson, D.-E. Box jellyfish use terrestrial visual cues for navigation. *Curr. Biol.* 2011; 21: 798–803. DOI https://doi.org/10.1016/j.cub.2011.03.054

[29] Garm, A., and Nilsson, D.-E. Visual navigation in starfish: First evidence for the use of vision and eyes in starfish. *Proc. R. Soc. B* 2014; 281: 1–8. DOI https://doi.org/10.1098/rspb.2013.3011

[30] Kirwan, J. D., Graf, J., Smolka, J., Mayer, G., Henze, M. J., and Nilsson, D.-E. Low resolution vision in a velvet worm (Onychophora). *J. Exp. Biol.* 2018; 221: 175802. DOI https://doi .org/10.1242/jeb.175802

[31] Nilsson, D.-E., Warrant, E. J., Johnsen, S., Hanlon, R., and Shashar, N. A unique advantage for giant eyes in giant squid. *Curr. Biol.* 2012; 22: 1–6. DOI https://doi.org/10.1016/j.cub.2012.02.031

[32] Wiederman, S. D., Shoemaker, P. A., and O'Carroll, D. C. A model for the detection of moving targets in visual clutter inspired by insect physiology. *PLoS One* 2008; 3: e2784. DOI https://doi.org/10.1371/journal.pone.0002784

[33] Parker, A. *In the Blink of an Eye: The Cause of the Most Dramatic Event in the History of Life.* London: Free Press, 2003.

[34] Zhao, F., Bottjer, D. J., Hu, S., Yin, Z., and Zhu, M. Complexity and diversity of eyes in Early Cambrian ecosystems. *Sci. Rep.* 2013; 3: 2751. DOI https://doi.org/10.1038/srep02751

[35] Nilsson, D.-E. Eyes as optical alarm systems in fan worms and ark clams. *Philos. Trans. R. Soc. Biol. Sci. B* 1994; 346: 195–212. DOI https://doi.org/10.1098/rstb.1994.0141

[36] Bok, M. J., Capa, M., and Nilsson, D.-E. Here, there and everywhere: The radiolar eyes of fan worms (Annelida, Sabellidae). *Integ. Comp. Biol.* 2016; 56: 784–795. DOI https://doi.org/10.1093/icb/icw089

[37] Bok, M. J., Porter, M. L., Ten Hove, H. A., Smith, R., and Nilsson, D.-E. Radiolar eyes of serpulid worms (Annelida, Serpulidae): Structures, function, and phototransduction. *Biol. Bull.* 2017; 233: 39–57. DOI https://doi.org/10.1086/694735

[38] Gehring, W. J., and Ikeo, K. *Pax 6*: Mastering eye morphogenesis and eye evolution. *Trends in Genetics* 1999; 15, 371–377. DOI https://doi.org/10.1016/S0168-9525(99)01776-X

[39] Nilsson, D.-E., and Arendt, D. Eye evolution: The blurry beginning. *Curr. Biol.* 2008; 18: R1096–R1098. DOI https://doi.org/10.1016/j.cub.2008.10.025

2 Visions

PAUL FLETCHER

> Thus, external perception is an internal dream which proves to be in
> harmony with external things; and instead of calling hallucination a false
> external perception, we must call external perception true hallucination.
> Hippolyte Taine, *De l'intelligence*, 1870 (trans. T. H. Kaye)

Introduction

Visual perceptions sometimes occur when there is no objectively identi-
fiable external stimulus that could account for them. These visions – or
visual hallucinations – come in many forms, from the simple floating
geometrical shapes that may herald migrainous attacks to the elaborate
and meaningful scenes that emerge in extreme states of perturbation:
states induced for example by psychedelic drugs, fever, or certain forms
of mental illness. The content of such visions can be terrifying, sometimes beautiful, and frequently highly salient and meaningful to the
observer. The seer of such visions may recognise them as distinct from
reality or may be convinced of their veracity, interacting with them and
seemingly oblivious to those around them. They may be transient and
mutable, morphing from one shape or form to another and lacking the
consistency, predictability, and persistence that most of us experience as
characteristic of the visible world. Conversely, they may be perceived as
reliable, predictable, and constant in their manifestation. For each of
these features, there is a continuum with different people experiencing
their visions as more or less complex, meaningful, convincing, and
constant.

A percept in the absence of an explanatory stimulus is the formal
definition of a hallucination, and the term vision is, in the clinical setting

at least, largely taken as a shorthand for visual hallucination. For many, the term hallucination has a predominantly clinical ring to it. They see it as the hallmark of madness: after all, to perceive – in any sensory domain – objects or occurrences that the majority does not is to be fundamentally separated from reality and, therefore, likely to think and behave in irrational ways. Such perceptual experiences, and the accompanying belief systems that emerge from and accompany them, have been considered as, almost by definition, beyond comprehension [1]. Moreover, given that we experience ourselves, and our relationships to others, largely through the acceptance of shared beliefs and models, the presence of these fundamental challenges to reality profoundly affects a person's capacity to fit comfortably into their social surroundings and the capacity of those around them to accept and empathise with them. The consequences can be, and frequently are, ostracism, isolation, and stigma. It is not always the case: a person who experiences visions may also be seen as blessed with a capacity to make contact with a deeper reality, perhaps a privileged experience of the divine or the arcane. However, this perspective too brings with it a tendency to isolate the visionary.

Here, I outline a different perspective of visual hallucinations, one that is rooted in a normative view of how the visual system processes the complex patterns of visible light, the small splinter of the electromagnetic spectrum to which we have conscious access and from which we strive to make sense of the visible world. I suggest that, when we examine the computational challenges that this system faces in doing this, along with the structural and functional principles that govern how it responds to these challenges, we see that normal, healthy vision can, at least in part, fit the formal definition of a hallucination. That is, the system creates perceptions that are not readily or entirely explicable in terms of an external reality. Paradoxically, this very fact may be at the heart of our capacity to comprehend and interact optimally with that reality. In simple terms, the system is an inveterate and automatic maker-up and filler-in. It presents to consciousness a representation of the causes of its sensory inputs, but it does so following extensive pre-processing. It is this automatic and pervasive pre-processing, I argue, that renders us liable to experience hallucinations: some useful and adaptive, some harmful.

Initial Caveat: Perception as a 'Controlled Hallucination'

This suggestion – that, even under normal circumstances, the realities of our vision (and other senses) are questionable – dates back at least to Hippolyte Taine (see the epigraph to this chapter) and is made with increasing frequency in cognitive neuroscience and related fields. It is worth stating at the outset, though, that, while I endorse it, I deny two attendant ideas that it sometimes ushers in. (i) I do not agree with the popular suggestion that this state of affairs compels us to accept that external reality is an illusion. Rather, it suggests to us that we should be mindful that our grasp on that reality necessarily entails some inadvertent creativity on the part of our perceptual apparatus. (ii) I would not argue that recognising the indirectness of our apprehension of reality entails accepting that we are mistaken or faulty in our interactions with the world. Frequently, demonstrations of the active, creative, and constructive, and therefore non-veridical, nature of vision rely on highly artificial stimuli that are deliberately engineered to produce a mismatch between expectation and actuality: we are amused to find that the compelling perception of a particular state of affairs is, on closer scrutiny, an illusion, for example, two objects that we firmly believed to be of different sizes are in fact the same, or the one that appears larger is in fact smaller, or that the line that we see as curved is in fact straight, and so forth. It does not follow that we are continually misperceiving and misinterpreting the world just because we are fooled in these isolated, and highly artificial, circumstances. It is important to remind ourselves that illusions tell us something very useful about our appreciation of reality, but they do not tell us that it is necessarily or invariably wrong. Rather they indicate that the perceptual strategy that we consistently employ can be, under certain circumstances, prone to error. But, as we move about the world, interacting fruitfully with it and using our expectations to shape perception and comprehension, there is no reason to suppose that we are in error and prey to the same misapprehensions. The world tends to be better behaved than the mischievous psychologists who construct these visual tricks.

Perception as an Active Process

Combining Sensory Input with Prior Expectations: Costs and Benefits

Our perceptual grasp on reality is a manufactured one. Vision does not entail a photographic record of the inputs to our eyes but rather a blend of those inputs with predictions or expectations that are related to our past experience as well as to the embedded structural and functional attributes of our visual system. This system works well in optimising our ability to comprehend and interact with noisy, intensive, ambiguous, and partial information about the true state of the world. But it comes with two costs, a small one and a big one. The small cost is a vulnerability to illusions that emerge when presented stimuli seem to, but do not, fit with our expectations. The larger cost is that the system seems ready-made to hallucinate – to create percepts in the absence of stimuli. These percepts may occur in the context of healthy processing, as, for example, in sensory deprivation, during which hallucinations may be strong and vivid. They may also occur in the setting of disturbances to our cognitive or neural responses, as, for example, with drugs or neuropsychiatric impairment.

By understanding the normal role and function of the visual system, and how it represents reality on the basis of its inputs, we may come to understand visions and to see them not as incomprehensible dislocations from reality and rationality, but as the products of a highly constructive and creative system that has been altered, unbalanced, or disturbed in some way.

In this chapter, therefore, I examine visual perception and consider the ways in which it may be perturbed. I shall begin by considering the idea, one that emerged in the 1940s and 1950s, in cybernetics and information theory, of the need for an agent to construct a model of its world in order to interact with it fruitfully and optimally. I then consider the necessity that perception is a process not merely of passive reception but of inference. Crucially, it is inference that draws on the predictability in the signal and requires that the organism carries within it some prior knowledge of the world: a prior knowledge that can be drawn upon to enhance efficiency and resolve ambiguity. I discuss the ways in which this principle – using prior knowledge to make predictions – may be enacted within the brain and I then apply these ideas to the central question: what might cause a person to experience visions?

Starting Ideas: Cybernetics, Models and Prediction

Before considering key principles of perception, it is worth thinking briefly and broadly about brain processes more generally. Across current cognitive neuroscience, one discerns a pervasive influence of the ideas of the mid-twentieth-century cyberneticists [2]. Although understanding the brain was not the principal goal of cybernetics, the researchers in this field were concerned with how systems controlled or responded to other systems, and their insights were readily applicable to brain function and dysfunction (indeed, a number of influential cyberneticists earned their living as psychiatrists).

The term 'cybernetics' was introduced by Norbert Weiner. It is derived from the Greek 'steersman', which captures the idea of a system or agent responding to changes in its environment – winds, tides, currents – in order to minimise the effects of those changes on the desired course of the boat. This leads to the question of the challenges that constructing such a model will face: notably the intensity and complexity, as well as the inherent ambiguity, of sensory signals that must be processed and interpreted.

Early cyberneticists built simple agents with a limited repertoire of responses to the external environment, observed their behaviours, and generated simple mathematical models to formalise the insights from these observations. Prominent among these was Ross Ashby's Homeostat [3], which comprised a set of linked units, each sensitive to (potentially perturbing) electrical inputs from the other units. A single unit would seek to regain its equilibrium, and, at the same time, this would generate outputs that would themselves perturb other units. While the details are not germane to the current discussion, the creation of this complex equilibrium-seeking device ultimately led Ashby to two related insights that are at the core of the current discussion. The first was 'the law of requisite variety', which is essentially that, for one system to exercise control over another (i.e. to respond to changes in that system in order to minimise the impact of those changes), it must have a repertoire of states at least as great as the number of states visited by the system that is being controlled. Thus, the number of requisite states in the agent has to match the number of ways in which its environment may perturb or influence it. Ashby's second insight, which has come to be known as the 'good regulator theorem', also relates to the prerequisites for one system to be able to control another. It is a mathematical formalisation of the observation that, for an agent (controlling system) to successfully mitigate the impact of changes in its environment (system to be controlled), it must be a model of that environment. As the classic paper by Conant and Ashby [4] states, '... any regulator that is maximally both successful and simple must be isomorphic with the system being regulated ... Making a model is thus necessary.' Moreover, Conant and Ashby pointed out that 'The theorem has the interesting corollary that the living brain, so far as it is to be successful and efficient as a regulator for survival, must proceed, in learning, by the formation of a model (or models) of its environment.'

While cybernetics has largely been superseded by other disciplines in the study of the brain, we should acknowledge its continuing implicit but widespread influence. The idea of the brain as a regulator, and therefore a model, of the world is powerful as it encourages us to consider perception, cognition, and action in the light of an over-arching purpose: to maintain the stability of the organism in the face of potentially survival-threatening changes. Moreover, it frames perception in terms of an attempt to recreate, in the form of an internal model, the associations and patterns in the world that give rise to the incoming signal. This idea, in the context of visual processing, will be developed in the next sections and becomes important as we try to understand the nature of the changes or disturbances that lead to visions.

Perception as Inference

Each of us lives within the universe – the prison – of his own brain. Projecting from it are millions of fragile sensory nerve fibres, in groups uniquely adapted to sample the energetic states of the world around us: heat, light, force, and chemical composition. That is all we ever know of it directly; all else is logical inference.

Vernon B. Mountcastle

An Important Distinction: Sensation versus Perception

Though sensation and perception are terms that may be seen as interchangeable, there are good reasons for making a distinction between them. I refer to sensation as the inputs from the world that initially impinge on the sense organs, signalling, as Mountcastle says, the energetic state of the world in a limited number of forms – heat, light, force, and chemical composition. I consider perception as the experience of that external reality. I should add that, when I say 'external', it is important to clarify that I mean external to the brain, since there are perceptions of our bodily states – collectively known as interoception – that form part of the experience (however, since this essay concerns vision, I shall not be further considering interoceptive processing).

In simple terms, sensations form the set of signals about the world and perception refers to how we apprehend the causes of those signals. Perception, as Mountcastle says, entails a process of inference, and the nature of this inference is worth scrutinising since it is germane to how we may understand visions.

The idea that perceiving the world is an active inferential process is an old one and, in early form, was elegantly outlined in the work of Hermann von Helmholtz (1821–1894), the German scientist whose work on perception provided a compelling early challenge to accepted ideas of perception as an essentially passive process of receiving signals and passing them on. Rather von Helmholtz considered vision as a reconstructive process. He put it as follows:

> Objects are always imagined as being present in the field of vision as would have to be there in order to create the same impression on the nervous mechanism.

In other words, the impression on the nervous mechanism (the 'sensation') prompts the imagination (or inference) of what object could have caused such an impression. It is noteworthy that, in his assertion, von Helmholtz treats perception as an act of imagination, an idea that I will not pursue here but one that should give us some pause. Perception entails a form of reverse inference in which the sense data must be decoded according to their likely origin. This, in turn, carries a key implication – that the constructive process inherent to perception must be driven not just by the signals of external origin but by an existing knowledge, based on experience, of what sort of thing could be the probable cause of a given pattern of signals. This is what the early-twentieth-century philosopher Charles Sanders Peirce referred to as an abductive inference – the inference from the (sensory) data to its likely cause. In the next section, I examine more closely the idea that perception is an inference about sensation.

Perception as Inference – the Importance of Prior Knowledge

Abductive inference in vision is problematic because there is never a single solution to the question that implicitly arises from von Helmholtz's question: 'given this observation, what is the explanation?' Inferring the external object that might have given rise to a complex pattern of visual input is challenging because any one of a practically infinite array of three-dimensional objects could give rise to a single two-dimensional representation on the retina. But it is possible to refine the question to ask about optimal abductive inference: specifically, 'given this observation, what is the most likely explanation?'

The answer does not lie in the sense data alone but in the expectations or prior knowledge of the observer. Given a set of competing explanations for a visual sensation, the rational inference depends on the prior probabilities of those competing explanations, As a toy example, if one hears hoof taps in an English country lane, the sensory signal may not distinguish between the possibilities that, around the corner, there may be a horse, a donkey, or a zebra. But prior knowledge in the rational perceiver would exclude the latter as being simply too improbable on the basis of what is already known about the current context. Similarly, a rapid movement of a dark shape at head height is more likely to generate the perception of a bird or bat than a cat. But if one were under water when confronted with these data, then none of those three possibilities would compete with the probability that the data are caused by a fish.

This is a rather dry description of a ubiquitous and fascinating process. It is more compellingly and interestingly demonstrated with a simple graphical example. On first looking at Figure 2.1(a), to most people it appears meaningless: it is a degraded image in which the black and white patches do little to convey its content. Even when told that it is an image

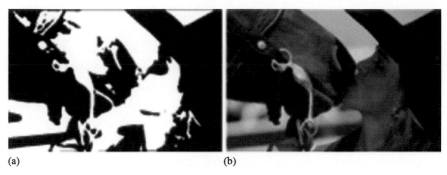

(a)　　　　　　　　　　　　　　　　　(b)

FIGURE 2.1 A demonstration of how knowledge aids perception by removing ambiguity. On first viewing image (a), few people can discern any meaning in it, experiencing it primarily as an incomprehensible collection of black blobs. After scrutinising image (b) (which depicts the original template image from which image (a) was derived, it becomes possible to discern the contents of image (a). As image (b) becomes more familiar, it becomes impossible not to see those contents – an effect that may last some time so that the contents of image (a) remain clear even when it is seen alone several weeks after seeing image (b).

of something and encouraged to look carefully in order to discern that image, one is hard-pressed to find meaning, though individual parts may perhaps resemble objects or figures. However, when provided with the relevant image that was used to create it, things can become much clearer. The clear image endows an observer with the prior knowledge that can then aid her in discerning the previously hidden contents of the degraded image. It may not be obvious immediately, and the observer may have to look from one image to the other a few times, but, once revealed, it becomes difficult not to see the contents of the previously unintelligible stimulus. Critically, the image itself has not changed, only the experience of the observer. This demonstrates the degree to which the perception of visual sense data is an active process that is based on prior knowledge. (See also [5].)

Another striking example of how knowledge or expectation can make sense of otherwise ambiguous or incomplete data, allowing a clear and informative percept to be created, is in the form of degraded writing, when either large parts of words are occluded or the majority of letters are rearranged. As a small experiment, try using a ruler or a sheet of paper to cover the bottom half of any line on this page. You may be surprised how effortless it remains to make sense of the words despite the fact that the bottom half of each has been occluded.

Interestingly, there is evidence that the value of prior knowledge as a means of reconstructing reality resides not just in its capacity to remove ambiguity but also in producing a change in sensitivity to stimuli that accord with expectations. This was demonstrated by Teufel and colleagues [6], who showed that a person's ability to detect a very faint, and otherwise imperceptible, contour is enhanced if they have prior knowledge about a similarly oriented contour that is not physically present but exists in the mind of the observer. This demonstrates that high-level comprehension of the contents of an image can alter the low-level visual processing of parts of that image.

There are many demonstrations of the influence of knowledge or expectation on our visual perceptions. Striking among these, as described above, is the wide array of visual illusions that rely, for the most part, on creating a tension between what a person predicts and what the sensory data objectively show. For example, illusions based on implied perspective, such as that shown in Figure 2.2, work by creating an expectation in the observer that one object is further away than another, leading to the illusion that two objects of objectively the same dimensions are of different sizes, because one is experienced, on the basis of the prediction induced by the perspective, as being further away than the other. If one object is further away than another but occupies the same physical space then it 'must' be the bigger object, and this creates the compelling perceptual experience of differently sized objects. This is demonstrated very simply by creating a perspective stretching into the distance and then placing identically sized objects at different points on this illusory perspective as I have done in Figure 2.2.

More on Prediction: A Key Process in Visual Processing

In the previous section, I have considered the brain as being engaged in the process of trying to model external reality in order to comprehend and manage it. I outlined some of the difficulties in this challenge, most notably, the fact that external reality communicates itself to us in a partial and ambiguous way. We have sensory apparatus that enables us

FIGURE 2.2 The influence of expectation on perception. Here, an expectation of relative distance is produced by the railway lines, which give the impression that the upper portion of the image is further away than the lower portion. When two identical objects (the cartoon trees) are placed in the upper and lower parts, one is experienced as being further away than the other, meaning that, since they are objectively the same size, this more distant tree must 'really' be larger. This tree is indeed perceived as being bigger than the one which, by virtue of the railway-line-induced perspective, appears closer to the observer.

to 'sample the energetic states of the world' as Mountcastle put it, giving us information about heat, light, force, and chemical composition, and, on the basis of this slender evidence, we must construct our model of reality. By using prior knowledge to generate expectations and to make predictions, we might more optimally engage in this process of abductive inference, allowing perception to emerge from an integration of prior expectation with incoming sense data.

This carries with it a possible implication that high-level abstract processes – knowledge and memory – are ubiquitous in visual perception, but this would be a simplification. Undoubtedly, our abstract understanding of how the world fits together, how objects associate with each other and with particular contexts, how particular features (e.g. feathers) are unlikely in particular settings (e.g. fish), gives us a powerful means of resolving ambiguity and making sense. But there are forms of prediction that lie much deeper, that seem to be embedded within the system and that act ineluctably and invariably from the instant that the signals from the world meet the earliest sensory apparatus. These forms of prediction, though they are automatic and fundamental, serve the same overall purpose as higher-level, more reflective predictive knowledge. They also seem to confer the possibility of optimally efficient signal transmission in a limited-capacity system that is being required to deal with an intensive and complex input.

The term 'predictive coding' was coined to describe these early operations [7]. While some might use the term interchangeably with the sorts of high-level predictive processes described above, it is perhaps more useful to think of predictive coding as being one particular form of predictive processing, a means of capitalising on environmental regularities in order to make useful predictions. In order to understand more readily the nature of predictive coding, it is useful to examine its antecedents in information processing theory and telecommunications before seeing how researchers have recognised similar operations occurring in the biology of the visual system.

Predictions at the Earliest Stages of Visual Processing – Removing the Predictable Signal

In an episode of the Simpsons, Homer sets his mind on becoming an inventor. His success is limited. One of his inventions causes particular distress to his family: The *'Everything's OK' Alarm* produces a painfully loud shriek every three seconds just as long as everything is OK. (While it is impossible to turn it off once started, it does, fortunately, break easily.) The irrationality of such a device, though lost on Homer, is recognised in the early visual system, as we see below.

Specifically, signals from the world are carried in the form of spikes of activity in neurons, and these spikes are expensive both in terms of the energy they use and in terms of the processing resources that they demand (the latter issue is especially problematic for a limited-capacity system trying to deal with a huge amount of signal). It makes sense, therefore, to find some way of coding information that enhances efficiency and reduces energy and computational cost.

This was a challenge faced by telecommunication engineers in the middle of the twentieth century; and their solution draws on a principle similar to that found in biological vision. Specifically, a television image demands the rapid transmission of a large amount of signal in a limited-capacity apparatus. The image itself contains multiple sub-units of varying image colour and intensity. While the corresponding colour/intensity values of each sub-unit could be transmitted one by one so that the information allows reconstruction of the image at the other end, this would be a laborious process and made even more demanding for a rapidly changing image, which would require the transmission of updated colour/intensity values as rapidly as possible. We might take shortcuts: for example, we might divide each successive image into fewer sub-units, leading to a reduced information load (i.e. fewer values to transmit). But this would come at the cost of a more uneven ('pixelated') picture with reduced spatial resolution and limited aesthetic appeal. Or we could choose to transmit images at a reduced rate (once every half-second, for example, rather than more frequently) but this would lead to jerky, unnatural movements.

The shortcut that played a big part in successfully meeting the challenge relates to designing a system that is able to capitalise on the predictability of signal [8]. Consider a film scene in which a small shape on the horizon moves slowly across a beach. Transmitting such a scene is made vastly more efficient by capitalising on the predictability of the scene itself both at any one time (large patches of sand, sea and sky are relatively homogeneous, each point serving as a good predictor of the colour and intensity of its immediate surroundings) and across time (even though the walker is moving, much of the rest of the scene is relatively constant: aside from some movements in waves, clouds, etc., it is the walker moving towards us that accounts for the most salient, frame-to-frame changes). So why transmit a single value for each sub-unit of the image and then re-transmit it for each separate frame when much of this information will be predictable and therefore redundant? To do so would be to make the error that Homer did with his *Everything's OK* alarm. It is much more efficient to identify and transmit areas of the image which are not predictable on the basis of their surroundings, for example, the borders between sea and sand or sea and sky. In focusing on what is unexpected – what is not predicted by its surroundings or what has gone before – we are dealing with fewer values and hence improving the efficiency with which the necessary information may be transmitted.

In 1982, in a landmark paper, Srinivasan and colleagues [7] made a compelling case that the principle described above is core to our visual apparatus. Within the retina (the light-sensitive multi-layered surface of the back of our eyes) there is an arrangement of cellular interaction that seems to cancel out predictable signals and emphasise unpredictable signals. This occurs, under good lighting conditions, through mutual lateral inhibition. Two neighbouring cells stimulated to the same degree will inhibit each other, cancelling activity out and meaning that no resultant signal is sent forward via the optic nerve to the brain. Thus, if two adjacent areas of an image have the same intensity (and thus activate neighbouring cells to the same degree), the system effectively blocks this signal – refusing to convey predictable information. Conversely, if neighbouring areas do not reliably predict each other, the retinal cells will not cancel each other out, and signal is conveyed for further processing by the system. This signal is essentially a mismatch between the expected and the actual signal – a quantity referred to as *prediction error* – and predictive coding refers to this process of coding a signal in terms of its unexpectedness: its prediction error.

Prediction within a Hierarchical System: Levels of Balance between Expectation and Input

The examples of integration between prior knowledge and current input given in the previous section highlight the potential value of such a system in allowing optimal inferences to emerge in the face of a very challenging set of data from the world. I have alluded to 'high-level' and 'low-level' instances of these prediction-based inferences, and now it is useful to consider how the same recurring processes may be stacked into a hierarchical system such that the output of any lower-level predictive processing loop serves as the input for the loop at the next level up. The inferences made at any given level serve as the predictions that shape signal processing at the next level down. This set of inter-linked loops, each performing the same sort of computation, leads, so the theory goes, to a system of message-passing within the brain that enables it to build models of the world that are able effectively to employ knowledge of many different types of regularity and, at the same time, are able to flexibly adjust and to update themselves (when there is a prediction error) through learning so that the model is the best it can be at any given time and is also able to respond adaptively should the world prove inconsistent. We can thus modify the above ideas a little and consider the brain as a *hierarchical predictive processing* device. As we shall see, the introduction of this slightly modified idea yields a rich seam of explanation in considering how visions may arise in different circumstances, some perfectly normal, some as a consequence of disturbance or pathology.

I now introduce the idea that this prediction-based inference is a motif that occurs repeatedly within a hierarchical system and that the system as a whole, by aiming towards a minimisation of the prediction error signal, can achieve the ultimate goal, as articulated by Ross Ashby and the other cyberneticists [2], of being a good model of the world.

According to this view, the brain is a system in balance, with the overall aim being to minimise the prediction error signal. This signal acts as a good marker for how well the model as a whole is doing because, while our access to external reality is limited and there is no prima facie way of knowing whether our inferences are correct, the act of making predictions offers us a powerful means of assessing just how wrong we may be. Simply speaking, if our prediction error is high, our model is wrong. And, up to a point, the more successfully we can make adjustments to reduce the prediction error towards zero, the more confident we can be that we are updating our beliefs in approximately the right direction (there are many caveats to this statement, but we may leave them aside for now). Simply put, a policy of minimising our prediction errors offers itself as a credible means of ensuring that we are maintaining a realistic contact with our world.

I have tried to represent this idea in Figure 2.3(a) and I discuss it below, returning to the original question concerning a deeper understanding of how and why visions might arise.

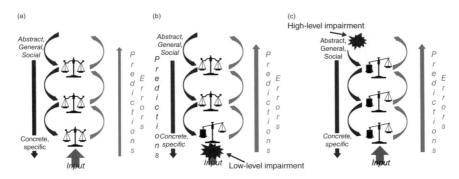

FIGURE 2.3 A simplistic representation of a predictive processing hierarchy as described in the text. (a) Overall, the system strives to balance and to minimise prediction error (unexpected signals) using knowledge- and experience-based expectations to balance inputs at multiple levels. Each level receives signals relating to prediction/expectation from regions above it in the hierarchy and input signals (possibly prediction errors) from the layers below it. The predictions are likely to vary in nature and quality at different levels of the system, e.g. being specific, physical, and concrete at lower levels and relatively general, abstract, and extending beyond the physical components of the environment (e.g. by encompassing social criteria) at higher levels. (b) An example of a disturbance to the system – Charles Bonnet syndrome (see the text). Here, an impairment to low-level input reduces meaningful visual input, meaning that expectations from the higher level are not counter-balanced by input. This leads to messages that are unduly weighted by internal expectation rather than external reality being passed up the hierarchy, i.e. visual hallucinations. Note that upper layers of the hierarchy continue to function normally and the patient can retain a good insight into the unreality of these visions despite their vividness and strong perceptual quality. (c) A (highly simplified) idea of how a very high-level perturbation (such as major psychological trauma) could act to produce a disruption in the experience of reality. Such a shift of high-level expectations about the world causes upcoming information to be interpreted differently at all of the levels below, from intermediate levels (e.g. interpreting facial expression as aggressive when it is in fact friendly or neutral) to interpreting unclear shapes as faces or agents. Thus, the high-level disturbance changes expectations that then change how the evidence from the world is sampled and interpreted, and this in turn seems to provide further evidence for the new – and potentially damaging and frightening – model of the world.

Summary: A System That Carries within It the Tendency to Create Visions

In summary, over the last two sections, I have outlined a simple principle that potentially meets the challenges faced by the brain in building a model of the world that is based on signals that are noisy, ambiguous, and intensive. It capitalises on the ways in which these signals are predictable and uses this predictability as a means of coding and transmitting them efficiently as well as decoding them in a way that enables optimal inferences about what is really out there. A key component in this process is the prediction error signal, which conveys information about deviations from expectancy. The key ideas were as follows.

1. The challenge of perception (in vision or any domain) is to make an inference about what is 'out there' that could account for the complex, ambiguous signals that are transmitted via our sensory apparatus. This sensory apparatus has a limited capacity, so the process must be as efficient as possible.

2. The transmission and interpretation of a sensory signal depends not merely upon the signal itself but on the experience and knowledge of the recipient. Perception is a process of inference that goes beyond mere guesswork by virtue of the predictions that a recipient can apply to the input.

3. Knowledge – and therefore prediction – comes in different forms and operates at different levels, from the basic predictability, across space and time, of visual scenes to the high-level, abstract, predictability that works in certain specific situations and contexts.

4. The system must incorporate these different levels of predictive knowledge and apply them in an integrated way to ensure that perception is optimally accurate and efficient. This has given rise to the 'hierarchical predictive processing framework'.

Reconsidering Visions in the Context of a Hierarchy of Predictions

I now consider how these simple ideas offer a framework for thinking about visions. To a large extent, this has been pre-empted by the discussion above on visual illusions. The tradition is to make a clear distinction between visual illusion and visual hallucination (the former being seen as a misperception of something that is there, whereas the latter is defined as the perception of something that isn't there). I disagree with this simplistic distinction, feeling that many illusions actually entail a creative or constructive process that is qualitatively indistinguishable from that which surely underlies hallucinations or visions. Indeed, in a remarkable nineteenth-century overview of hallucinations (*Phantasms of the Living*, by

Gurney, Podmore, and Myers), it was commented that 'illusions are merely the sprinkling of fragments of genuine hallucination on the background of true perception'. It seems reasonable to consider the visual system as set up in such a way that it is vulnerable to numerous possible imbalances between predictions based on stored knowledge or experience and current sensory inputs. In the case of visual illusions, these imbalances are artificially created such that the system is presented with a stimulus that both creates an expectation and simultaneously violates that expectation. With hallucinations, a comparable imbalance or violation occurs, but its cause is embedded in the system itself and may arise as a consequence of many different disturbances afflicting the system at different levels and in different ways. Applying the framework above provides a useful approach to examining how different alterations may lead to different forms of imbalance. Although the end result of each of these forms of disturbance may be broadly defined as visions, or visual hallucinations, the underlying differences in how the phenomena arise will ultimately prove crucial to truly understanding them and to attempting to manage or treat them.

Disturbing the Balance (i): A Change in Low-Level Sensory Input – Charles Bonnet Syndrome

"I often see big cats like tigers and panthers . . . sometimes inside the house but also outdoors. The images used to be scary . . . but now I can at least account for them." (First-person account of one of the hallucinations of Charles Bonnet syndrome – see www.charlesbonnetsyndrome.org)

In his 1760 *Essai analytique sur les facultés de l'âme*, Charles Bonnet described the case of his grandfather, who, in association with a growing impairment of vision caused by corneal degeneration, began to experience florid and elaborate visions of humanlike figures, birds, and buildings. This experience of visions occurring in the context of visual sensory impairment has been described many times subsequently, becoming known as Charles Bonnet syndrome. It is characterised by often very elaborate visions that are perceived as real and may, initially, cause profound distress to the sufferer who is both puzzled by the existence of these visions and frightened by their content. Interestingly, however, and in contrast to the hallucinations that accompany serious mental illnesses, the patient may come to realise, quite quickly, that these visions do not reflect reality and are indeed produced by their own mind. Thus, while the phenomena persist and may remain perceptually vivid and elaborate, they preoccupy and distress the person less and less with time, as is demonstrated in the quote above.

The Charles Bonnet syndrome seems highly paradoxical: why would a reduction in the strength or precision of visual inputs, caused by deterioration in primary sensory apparatus, be associated with such vivid visual experiences? In fact, we find a number of other instances in which attenuated visual inputs lead to this effect. For example, subjects in sensory deprivation chambers, in which input to all sensory modalities is markedly reduced, experience the rapid onset of hallucinations. And in situations where a person is confined for periods in a darkened environment, they may rapidly come to see flickering lights and shapes, geometrical patterns, and kaleidoscopic images. This can happen when people are placed in solitary confinement; hence it has become known as *the prisoner's cinema*.

This finds a simple explanation within the predictive processing framework. See Figure 2.3(b). If, under normal circumstances, predictions based on stored knowledge are integrated with sensory input, then an occlusion of that input leads to a perturbation such that expectations at various levels within the system (from low-level expectations of contours and shapes to higher-level ones relating to objects and scenes) are not balanced by sensory inputs from external reality. This imbalance could result in visual perceptual experiences being comprised primarily of those the brain creates, unconstrained by the world. Under such circumstances, even a small, but erroneous, prediction arising from higher in the system (e.g. of a contour or object) could be allowed an undue weighting since there is no reliable upcoming sensory signal to challenge it.

Importantly, because the underlying cause of this imbalance is a well-circumscribed, low-level impairment, the system as a whole should be able to function normally at higher levels, meaning that, even though the perceptual experiences may be convincing and vivid, other sources of information, such as the improbability of seeing tigers in one's home, the fleeting nature of the experiences (and the fact that they would not be corroborated by evidence from other senses), and reassurance from family and friends that these visions do not reflect reality, can allow the person to treat them as the products of their minds rather than a true reflection of the real world. This is indeed the case, as is captured by the quote above, though the sufferer may go through difficult and unpleasant phases of finding it hard to tell visions from reality.

Disturbing the Balance (ii): A Shift in Expectations Producing a New Reality

These are men whose minds the Dead have ravished.
Memory fingers in their hair of murders,
Multitudinous murders they once witnessed.
Wading sloughs of flesh these helpless wander,
Treading blood from lungs that had loved laughter.
Always they must see these things and hear them.

Wilfred Owen, Mental Cases

One of the striking observations relating to psychotic illnesses (illnesses involving an apparent loss of contact with reality and being characterised by, among other things, hallucinations) is that they are significantly associated with a history of trauma to the individual. In this respect, post-traumatic stress disorder (PTSD) presents a powerful example. It is not, strictly speaking, a psychotic disorder, but is characterised by 'flash-back' episodes in which the sufferer vividly relives experience of the original trauma: its terrors, sights, sounds, smells. They may have the profound sense of being back at the scene of this trauma and, in this sense, they may be truly dissociated from current reality. Such psychotic-like episodes can be triggered by an incidental and perhaps indirect reminder of the original trauma, even a simple word or phrase that is enough to reactivate the memory and to plunge the person into the past. The power and horror of the experience is captured by the lines above from Wilfred Owen, written in hospital only a short while before he returned to meet his death in the trenches in 1917. There are also an appreciable number of instances where a person has experienced earlier traumas and, although they don't show the clear-cut symptoms of PTSD, may experience intermittent periods when they experience altered beliefs and perceptions of the sorts described above.

I would suggest (see Figure 2.3(c)) that it is useful to consider these psychosis-like episodes in terms of a profound, and perhaps long-standing, change in the person's model of the world, one that occurs as a very understandable response to the earlier traumatic experiences. Someone, for example, who suffered neglect or violence at an early age is likely to have established a very different perspective on what they should expect from other people and how they should view themselves in relation to others. Unpredictability, uncertainty, and the expectation of hostility or aggression will profoundly influence how expectations shape ambiguous sensory stimuli. It follows that these high-level changes in expectation could mean that an input that might more commonly otherwise be perceived as neutral or non-salient would be seen as threatening or disturbing. One can see such effects in a milder form, as part of normal experiences: a nervous person walking in darkness on an empty street may become convinced that an inanimate object is an assailant lying in wait. Or a self-conscious person who hears laughter on entering a room may perceive it as derisory and cruel. Facial expressions are often difficult to interpret, and how we view and interpret them tends to be powerfully influenced by our beliefs about the motives and intentions of the person we're looking at.

These examples illustrate the key point that our higher-order beliefs can influence how we sample and interpret our sensory inputs. A hierarchically organised system that uses past experience and knowledge to generate inferences about an ambiguous world will create deeply personal interpretations. Depending on one's personal history, these interpretations may be far-removed from those that might come more naturally to others with very different histories and, correspondingly, different sets of expectations. Crucially, once an inference has been made – for example, that someone's facial expression signals malevolent intent – this serves as further evidence in favour of the belief that generated this inference. And so the system acts in ways that find evidence to support its predictions. While, ideally, beliefs should be updated when evidence does not support them, they may find ways to persist because we naturally interpret or sample evidence in particular, and often self-fulfilling, ways.

Disturbing the Balance (iii): A Pervasive Change in How Input and Expectation Are Integrated

The predictive processing framework relies on a complex, neurochemically mediated system of message passing. Though the precise mechanisms of this system remain largely unknown, there are certain key neurotransmitter systems that may be key. I will not go into detail here, but, for the present purposes, one can envisage another alteration to the system that arises not from impairment of its input ((i) above) or a shift in experience-dependent predictions ((ii) above) but from a disturbance in the message-passing connections, perhaps as a consequence of alterations in the neurotransmitters or their receptors. Such alterations could arise from effects caused by drugs, particularly the so-called hallucinogenic drugs such as LSD and psilocybin, or from prolonged use of amphetamines (which may disrupt function of the transmitter dopamine, and produce effects that partially resemble mental illness such as schizophrenia, itself thought to involve altered dopamine function).

To give one example, suppose, as has been suggested in relation both to schizophrenia and to amphetamine-induced psychosis, excessive activity of dopamine is associated with altered beliefs and perceptions. How might the low-level, neurochemical change produce such powerful changes in subjective experience? One suggestion is that dopamine is crucial to prediction error signalling (see above for a description of the importance of prediction error in the current framework). An elevated level of dopamine signalling could therefore generate a subjective sense that one's existing predictions are wrong, that the world is no longer adequately accounted for by one's previously successful model. Over the course of time, this persistent and inexplicable change would drive the individual to seek new explanations and new accounts. These might become more outlandish in their explanatory scope, perhaps introducing ideas that the person had previously rejected as being improbable or incredible. Perhaps all of these strange new experiences might be accounted for only by a nefarious agent seeking to confuse or harm the person. This in turn would produce further imbalance in the system such as those described in (ii) above, with the result that the new beliefs changed interpretations and inferences, thereby creating new evidence that seemed to support them.

This is highly speculative, but serves to show how a specific alteration in the nature of neural communication within such a predictive system could lead to a profoundly and disturbingly altered model of the individual's world and their own place in it.

Conclusion

Visions – and hallucinations in general – present a challenge to our understanding. How do we perceive things that objectively aren't present? I have tried to show how a relatively simple framework, based on the idea that the brain is engaged in prediction-based inferences about the causes of its sensations, can help us to begin to understand these baffling and complex experiences. Such experiences can be characteristic of psychiatric or neurological illness but might also occur under normal circumstances. Importantly, the framework inspires an understanding that is not purely biological but also draws on other levels of explanation, including one's personal history and social interactions. This reflects the important observation that the brain's role is to model its world. Given the centrality of this idea, a comprehensive account of brain function cannot possibly ignore the wider environment in which the brain exists and functions. Nor, conversely, should we focus purely on social or environmental factors without considering the brain that is engaged in making inferences about these factors and thereby profoundly shaping the individual agent's experience and interpretation of them.

A huge amount of work remains to be done if we are to use this framework to generate truly useful insights into the nature of visions and, where needed, their treatment. But its embodiment of a simple and powerful idea – that we are continually making inferences by means of integrating prediction with input – offers fascinating perspectives to the psychiatrist, or to anyone interested in brain function and dysfunction.

References

[1] Jaspers, K. *General Psychopathology*. Manchester: Manchester University Press, 1962.
[2] Pickering, A. *The Cybernetic Brain: Sketches of Another Future*. Chicago, IL: University of Chicago Press, 2010.

[3] Ashby, W. R. *Design for a Brain: The Origin of Adaptive Behaviour.* London: Chapman and Hall, 1960.

[4] Conant, R. C., and Ashby, W. R. Every good regulator of a system must be a model of that system. *Int. J. Systems Sci.* 1970; **1**(2): 89–97.

[5] Teufel, C., Subramaniam, N., Dobler, V., Perez, J., Finnemann, J. et al. Shift toward prior knowledge confers a perceptual advantage in early psychosis and psychosis-prone healthy individuals. *Proc. Natl Acad. Sci. USA* 2015; **112**(43): 13401–13406.

[6] Teufel, C., Dakin, S. C., and Fletcher, P. C. Prior object-knowledge sharpens properties of early visual feature-detectors. *Sci Rep.* 2018; **8**(1): 10853.

[7] Srinivasan, M. V., Laughlin, S. B., and Dubs, A. Predictive coding: A fresh view of inhibition in the retina. *Proc. R. Soc. Lond. B Biol. Sci.* 1982; **216**(1205): 427–459.

[8] Harrison, C. W. Experiments with Linear Prediction in Television. *Bell Syst. Tech. J.* 1952; **31**(July): 764–783.

3 Colour and Vision

ANYA HURLBERT

When Turner daubed a red buoy in his seascape *Helvoetsluys* (Figure 3.1), what did he mean? For Turner, the red created contrast, and, in making that mark, he meant to generate salience and arouse interest, to dominate his rivals and draw in his admirers. He painted in the red blob after the painting had been hung, a response to Constable's *Opening of Waterloo Bridge*, which hung opposite in golden glory. Of the latter, the contemporary critic Charles Robert Leslie said 'the intensity of the red lead, made more vivid by the coolness of his picture, caused even the vermilion and lake of Constable to look weak' [1]. Turner's deployment of colour illuminates principles of colour perception, even if he did not consciously articulate them as vision scientists would today, as well as, most importantly, the meaning of colour in human life.

One may argue that Turner transformed from a painter bound by line and form into one consumed by colour as his life evolved from its troubled but hard-working beginnings into the greater freedom brought by wealth and respect. As a young man, engraving was natural to Turner, since he viewed colour largely in terms of 'light and shade' [1]; its value for him lay in the varying luminances of different colours. Monochrome lines and textures fitted his early drawings of narrative scenes and architecture. As he aged, his use of colour became freer, more abstract, and his work became more emotive and expressive. This evolution is apparent in his treatment of favourite sites which he drew for his Liber Studiorum, an early book of engravings (1806–1824), and painted again and again over his lifetime, adding colour and removing line. Norham Castle in Northumberland is a prime example.

Turner's interpretations of Norham Castle evolve from a drawing in his north of England sketchbook, 1798, all lines, through his mezzotint

FIGURE 3.1 J. M. W. Turner. *Helvoetsluys; – the City of Utrecht, 64, Going to Sea*, 1832. Oil on canvas. Tokyo Fuji Art Museum. Credit: Tokyo Fuji Art Museum, Tokyo, Japan/Bridgeman Images.

published in 1816, to a watercolour in 1823 (Figure 3.2). Then, in the early 1840s, after a 'long period of introspection' [2], 'in which Turner reminded himself of early subjects by looking again at Liber Studiorum prints and by having some of them reprinted', Turner returned to Norham Castle and painted a diffuse atmosphere of blue, brown, gold, and grey, 'which despite its deserved fame as an image, is nowhere on this planet, just a blue rock and an unearthly cow' (Figure 3.2). He began with classical depictions of a remote fortress and ended with ephemeral evocations, expressing the inability to capture contours through the haze of memory, but also emphasising their unimportance, and even their obstructiveness, in conveying the feeling of the place: its remoteness, its isolation, and its unity with nature, in the merging of the building into the sky, and of the animals with the earth.

(a)

(b)

FIGURE 3.2 Joseph Mallord William Turner. (a) *Norham Castle on the Tweed*, from Etchings and Engravings for the Liber Studiorum, 1816. (b) *Norham Castle – Sunrise, c.* 1845. Oil paint on canvas. Tate, London. Photos © Tate.

The evolving balance of line and form vs. colour and light in Turner's work reflects the debate on *disegno* vs. *colore* in art history, which in turn mirrors the divide between form and colour in the scientific approach to the understanding of human visual perception. As Leonardo da Vinci (1492) asked, 'Which is of greater importance? Form or Colour? That form should abound in beautiful colours, or display high relief?'[3]. Leonardo's Burlington House cartoon delineates his answer (Figure 3.3). The Virgin Mary sits on the lap of her mother, Saint Anne; the Christ Child blesses his cousin Saint John the Baptist. Nothing but charcoal and wash, contours and shading, is needed to convey the illusion of depth, curvature, and pose, and from these, spiritual relationships and enactment.

FIGURE 3.3 Leonardo da Vinci. *The Burlington House Cartoon*, 1499–1500. Charcoal (and wash?) heightened with white chalk on paper, mounted on canvas. © The National Gallery, London. Purchased with a special grant and contributions from the Art Fund, The Pilgrim Trust, and through a public appeal organised by the Art Fund, 1962.

Colours, which 'honour only those who manufacture them', do not intrude on the serenity and grace of the figures.

The French academicians of the eighteenth and nineteenth centuries also diminished the significance of colour in paintings, arguing that it was too base, too sensual, and too mutable to convey deep truths. They moved away from statements about wealth of pigments and symbolism of colour to a pre-eminence of form, line, and shape. Ingres (1780–1867) argued that a *'noble contour'* 'may atone for other deficiencies' [4]. Delacroix (1798–1863), leading the charge for colour, retorted '[Ingres] has not the slightest inkling that everything in nature is reflection and that colour is essentially an interplay of reflections' [3].

In vision science, it has long been an aim to determine the importance of colour for perceptual and cognitive needs. The purpose of vision is often described in textbooks as to 'determine *what* is *where*', to identify objects and assess their physical properties, spatial location, and direction and speed of movement, in order to enable behavioural interactions with them. With respect to determining *what* an object is, one repeatedly addressed question echoes Leonardo's: which is more important in driving recognition, shape or colour? Most scientists would say that shape wins, but the answer is complicated, and the debate continues.

Prominent 'edge-based' theories of visual object recognition, for example, give primacy to geometrical features such as contour and shape [5–7] and treat colour as being secondary to these in signalling object identity [8]. Line drawings of objects such as forks, cameras, and mushrooms are recognised as rapidly as colour photographs [8], in support of these theories. A counterargument is that colour is not a defining feature of such objects, whereas shape is. Indeed, when colour is highly diagnostic, as it is for red strawberries, green avocados, and blue kingfishers, naming objects or classifying scenes is significantly faster when they are presented in their typical colours compared with atypical colours (e.g. blue pumpkins or pink forests) or entirely in greyscale alone [9–11]. Even for human-made objects without diagnostic colours, as for buttons, bottles, and stools, colour in fact speeds up object naming, when tested comprehensively [12]. Colour therefore acts not only in subordinate-level recognition, distinguishing limes from lemons, but also at a basic level, telling apart elephants from bananas.

Such studies imply that colour is integrally entwined with shape in the neural representations of objects [11, 13]. It has yet to be shown that colour alone may directly arouse complex visual memories of objects, independently of shape. It is true, though, that colour assists shape even earlier along the route to recognition, through delineating edges better. Colour is a critical cue for image segmentation, and is more effective than luminance in demarcating meaningful material changes in favour of incidental illumination effects.

In the image of a fruit market in Figure 3.4, there is shadow and bright sunlight. The luminance image captures all these boundaries, but the chromatic image shows the objects only. This idea underpins computer algorithms for automatically segmenting images before applying object-recognition routines (e.g. [14, 15]).

In finding material boundaries [16], colour helps to distinguish shadows from stains, or specular highlights from flecks of bright paint, as in Heda's *Still Life with a Gilt Cup* (Figure 3.5).

FIGURE 3.4 A fruit market. (a) Full colour photograph. (b) Luminance information only. (c) Chromatic information only. The original image (a) is reproduced by permission of the Minolta Corp., via Andrew Stockman.

FIGURE 3.5 Willem Claesz. Heda. *Still Life with a Gilt Cup*, 1635. Oil on panel. The Rijksmuseum, Amsterdam. Note the multiple instances of specular highlights on glass, gilt, pewter and silver. Their chromatic signatures and luminance profiles distinguish them from surface markings.

In providing a distinctive segmentation cue, colour also speeds up visual search, enabling red Ts to pop out quickly in an array of blue Ts [17]. It enables us to find *that* book on the shelf or rental car in the lot, solely by their colour. Colour not only segments the image into individual parts, and thereby directly assists processes of object recognition, but also provides additional information about the physical properties of those parts, in particular, about the material from which they are made. Material information in turn feeds into the assessment of other inherent qualities of the object – for example, edibility.

Colour, in everyday life, thus does many things for us, visually – making unconscious image processing faster and more efficient – and cognitively – providing cues to object identity, categories, labels, and physical properties. But colour also seems to have a direct route to the emotions, and I suggest that Turner turned to colour to express emotion more profoundly, not just to represent a clearer form of reality. Carl Jung, the Swiss psychoanalyst, placed colour not only at the wellspring of emotions, but also at the core of the self, unifying consciousness, personal unconsciousness, and the collective unconsciousness with its inherently concentric nature [18]. Colour is central to Jungian mandalas, the symbol of the self, and, throughout Jung's writing, colours represent feelings, archetypes, and 'psychic functions' [18]. Why colours should have such emotional heft is still a matter of scientific debate, but one explanation is that colours derive this power from their biological significance, which in turn follows from the material properties that they signal [19, 20].

The spectrum of light reflected from objects depends on the physical constituents of their surfaces: the type and concentration of the pigments they contain, the substrate in which the pigments are embedded, and the geometrical structure and transparency of the layers. These physical characteristics in turn depend on qualities such as freshness, ripeness, age, moisture, and chemical content. As a biological signal, colour may therefore indicate the edibility of fruit, healthiness of skin, viability of a mate, or toxicity of a toadstool. It is plausible that colour acquires its meaning from these natural sources. People might naturally find red arousing, attention-grabbing, because, over the course of human evolution, red has signalled either danger or delight – poison (the fly agaric) or passion (the flush of a monkey's bottom). Red spurs the viewer to action. In contrast, blue, being like the sky or distant mountains, encourages 'contemplation' and 'draws us after it', said Goethe [21]. Turner chose red to make his mark in *Helvoetsluys*, perhaps, in order to create that tension, accosting the viewer with a call to action in his sea of contemplation.

Yet, although the meaning of colour might arise from the biological significance of objects which naturally express colour, the quality that enables colour to convey meaning in itself is its abstractability. Conceptually, colour may be detached from objects. We are able mentally to represent and describe brilliant greens or light lilacs, without tying them to spring foliage or bearded irises. Although, in normal vision, colours remain within surfaces, the fact that, in some brain disorders, 'colours may seem to extend far beyond the confines of the object to which they belong' [22] suggests that the detachability of colour is not only conceptual, but also perceptual. At some level of visual processing, the neural representation of colour must be separable from objects. Recent neurophysiological and anatomical evidence indeed supports a duality in colour processing.

The standard view has long been that neurons signalling colour are segregated from those signalling form, beginning in the earliest stages of cortical processing (primary visual cortex or V1). Direct interrogation of neuronal mechanisms in non-human primates reveals that cells respond preferentially *either* to particular patches of colour, presented in the particular spatial locations to which they are tied via their retinal connections, *or* to particular orientations of luminance edges, but not both [23]. Colour cells cluster together in blobs in V1, and orientation-selective cells in inter-blob regions. In higher visual areas, their respective processing recipients remain segregated. Yet there is also increasing evidence for concomitant regions of integration [24]. Neuroimaging studies in humans suggest that cells simultaneously selective for colour and form are dispersed throughout early visual cortical areas. For example, small regions of V1 are highly selective for particular conjunctions of colour and shape – e.g. red counter-clockwise pinwheels [25]. And even in higher cortical areas involved in later stages of visual processing, where congregations of cells express exclusive preferences for highly complicated stimuli such as specific faces or places, there are cell clusters interested almost solely in colour [26], regardless of object shape or identity [27]. These findings support the idea that the visual system calls on colour not only as a simple feature to mark out surfaces of potential interest but also as a high-level, potent attribute on the fast track to meaning.

It is this separability that makes possible the debate between *disegno* and *colore* in art, and between form and colour in vision science. It also allows colour to be lifted from surfaces where it naturally occurs, and to carry and confer its emotional meaning to other objects. We humans have therefore naturally learned to release colour from form, to celebrate it as an independent entity, and to use colour as an adaptable label or symbol for more abstract concepts or categories. In human-made objects, for example, colour may instruct behaviours (at traffic lights signs or

recycling bins) [28] or symbolise political allegiance or group member-
ship. And, as for natural objects, artificially applied colour may therefore
signal danger or delight. In the gang warfare of 1990s South Central Los
Angeles, where the Crips wore blue and the Bloods wore red, 'you could
literally lose your life for wearing the wrong colour in the wrong place',
says Aqeela Sherrills [29].

In paintings, colour may serve as a symbol for either good or bad,
heaven or hell. But it might be that in painting colour acquires meaning
through another route entirely, not only through the artist's deliberate
tapping into naturally evolved or culturally imposed associations, but also
through the physical processes of painting itself. The pigment itself makes
the meaning of colour. When we thrill to the expansive blue of
Sassoferrato's *Virgin in Prayer* (Figure 3.6), a colour not seen in the surfaces
of natural objects, are we responding to its culturally symbolic purity and
heavenliness? Or have we learned the emotion from the ultramarine
pigment itself, its celestial richness deriving directly from its very expensive earthly origins, lapis lazuli in arduously reached mountain mines?

What Is Colour?

Until now, I have talked of colour as if it is a certain thing, known and
understood in the same way by all of us. Yet colour is neither a certain
object nor a physical property of objects. Instead, it is a perceptual
phenomenon, constructed by neural processes acting on the responses
of retinal receptors to light that enters the eye.

The fact that the colours you see are made by your brain, not 'out
there' in the world, is illustrated by the lilac chaser illusion [30]
(Figure 3.7). The animation begins with a ring of twelve lilac dots centred
on a small black cross, evenly spaced. The dots have blurry edges, and are
roughly the same brightness as the grey background. In succession, each
dot briefly flashes off, the other dots remaining static. If you fixate on the
cross, you will soon see a *greenish* blurred dot moving clockwise around
the ring, obliterating each lilac dot in succession. The lilac dots might
then disappear entirely, leaving a solitary green dot tracing out a circle.

The green dot is not there. That is, there is no frame in the animation
in which the green dot has been drawn in or otherwise specified; the only

FIGURE 3.6 Sassoferrato. *The Virgin in Prayer*, 1640–1650. Oil on canvas. The Virgin's cloak is painted in ultramarine, made from lapis lazuli. © The National Gallery, London. Bequeathed by Richard Simmons, 1846.

RGB values in the image correspond to the grey of the background and the lilac of the dots. The green dot arises from active processes in your brain. Its presence is a specific, and particularly powerful, illustration of the general phenomenon of afterimages caused by chromatic adaptation, of which many other examples exist. It also drives home the fact that seeing is subjective and probabilistic. This idea, that visual images are ambiguous and vision is an interpretive process, in turn has a long history, cutting across philosophy, psychology, physics, and biology, from Democritus to Locke, Goethe, and Newton, and is now the driving concept in much of contemporary vision science [31].

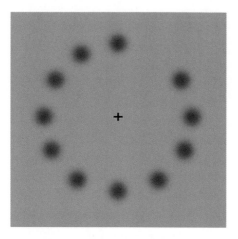

FIGURE 3.7 The lilac chaser illusion. The image is a still from the animation. See the description in the text.

Yet, people experience colour as an inherent property of objects: strawberries are red, lemons are yellow, my favourite blazer is teal. There is a dichotomy between the objectivity and subjectivity of colours: the objects seem to possess these colours, whereas, in fact, I possess mine and you yours. The polymathic scientist Hermann von Helmholtz said: 'Colours have their greatest significance for us in so far as they are properties of bodies and can be used as marks of identification of bodies' [32]. In other words, colours are intrinsic components of objects, and thereby serve as means to recognise them. Yet, at the same time, Helmholtz continued, we cannot see these intrinsic components except by 'eliminating the differences of illumination by which a body is revealed to us', through physiological processes [32]. In other words, Helmholtz implied, colours are indeed 'out there', but can be revealed only through the workings of the brain. And these, because of imperfections, may yield only approximations of the true physical state of the world.

Colour Constancy

Helmholtz's reference to 'differences of illumination' points to the critical feature in understanding colour perception: colour constancy. Colour

(a)

(b)

FIGURE 3.8 Claude Monet. (a) *Rouen Cathedral, The Portal and the Saint-Romain Tower, Full Sun*, 1893. Oil on canvas. Musée d'Orsay, Paris. (b) *Rouen Cathedral, The Portal and the Saint-Romain Tower, Morning Effect*, 1893. Oil on canvas. Musée d'Orsay, Paris. Photos © Musée d'Orsay, Dist. RMN-Grand Palais/Patrice Schmidt.

constancy is a perceptual phenomenon, by which people perceive object colours as staying stable under changes in illumination. It is generally agreed to arise from multiple mechanisms hard-wired into the workings of the visual system on multiple levels. It is unusual for an individual to be able to switch off colour constancy, although one could argue that Monet did. In his series paintings, Monet captured the radical changes in light spectrum reflected from objects under changing illumination, effectively disabling colour constancy. He paints the western façade of Rouen cathedral in full sunlight with rich golden hues; the stone is bright, etched in detail, standing proud against a darker blue sky. In early morning, he paints the cathedral, from almost the same vantage point, in dark, shadowy blues and violet, against a brighter sky tinged with yellow (Figure 3.8). In situ, to an ordinary human, the stone would appear beige,

FIGURE 3.9 Uluru, or Ayers Rock, in central Australia. Photographs taken in the evening over several minutes in July 2007. Photos © Anya Hurlbert.

less variable, more neutral. Monet painted colours of light of which we ordinarily would not be aware.

For ordinary humans in the natural world, failures of colour constancy seem rare. But Ayers Rock, or Uluru, in central Australia, provides one

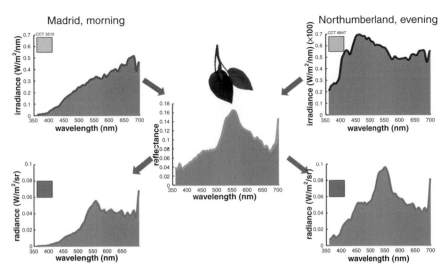

FIGURE 3.10 Colour constancy. Centre panel: the surface spectral reflectance function of a leaf. Top row: the spectral irradiance of daylight illumination, in the morning, Madrid (left) and in early evening, Northumberland (right). The inset squares show a neutral surface ('white') under corresponding illumination. Bottom row: the spectral radiance of light reflected from a leaf under corresponding illumination (Madrid, left; Northumberland, right). The inset squares show a simulated leaf surface.

example (Figure 3.9). Over the course of several minutes at sunset, the monumental sandstone formation appears to change in colour, from bright orange through deep russet to dark purplish-brown. Instead of attributing these changes to the changing illumination as the sun sets, the visual system seems to conclude, unconsciously, that the rock itself is changing. The lack of other reference surfaces in the otherwise vast empty horizon might contribute to this interpretation.

Colour constancy is often cited as a prime example of a perceptual constancy, in which the visual system transforms the varying two-dimensional image at the back of the eye into a stable representation of the three-dimensional world. Size constancy, another example, compensates for the decreasing retinal image size of a person as he or she moves further away, keeping his or her perceived height at six foot tall, instead of shrinking rapidly.

Colour constancy requires the visual system to compensate for the changes in the retinal image caused by changing illumination [33]. The spectral reflectance function of a surface describes the proportion of incident light the surface reflects at each wavelength (Figure 3.10). This property is intrinsic to the surface, being dependent on its physical and chemical composition. For example, the chlorophyll pigment in leaves absorbs light at short wavelengths and reflects light at middle and long wavelengths. The spectrum of light it reflects depends on the amount of light that falls on the surface at each wavelength, and therefore will vary with the illumination spectrum. At dusk, when the sun sinks, daylight contains more short-wavelength light than it does in mid-morning under a cloudless sky. Then, when sunlight dominates, the leaves reflect more middle- and long-wavelength light. Yet, because of colour constancy, the leaves do not appear to change from yellowish-green to blue during the day, but remain green. The visual system 'corrects' for the change in illumination spectrum.

In Helmholtz's view, this 'correction' is performed partly as an act of judgement, in order to reveal the inherent 'body' of the object. Contemporary computational models formulate colour constancy in a similar way, treating it as a problem of recovering an estimate of the surface spectral reflectance function from the reflected light spectrum, in which it is entangled with the illumination spectral power distribution. The problem is ill-posed, in that there is not enough information in the light signal to uniquely specify one without knowing the other, so the problem can be solved only by applying constraints or making prior assumptions. The underlying aim is, though, to recover a surface descriptor that remains invariant under changes in the illumination power spectrum, and which would ideally correspond to the constant colour that people perceive and, in turn, enable colour to signal material properties and object identity. In the historical debate over the nature of colour, some philosophers expressly argue that surface spectral reflectance *is* colour [31].

The Uncertainty of Vision: An Interpretive Process

Colour constancy is thus not only a cornerstone of visual perception but also a key exemplar of the 'inverse optics' approach to modelling vision [34, 35]: the idea that the human visual system inverts the optical process of image formation, reconstructing a three-dimensional world

FIGURE 3.11 Johannes Vermeer. *The Milkmaid, c.* 1660. Oil on canvas.
The Rijksmuseum, Amsterdam.

from a two-dimensional image. Vision scientists have learned from
artists' deployment of paint colour and texture over a flat surface which
image cues are most effectively exploited by the human visual system to
recreate the world. Vermeer's *The Milkmaid*, for example, represents a
scene rich with different materials – bread, ceramics, woven straw, metal,
wood, cloth, milk – as well as concentrated action, light, and space
(Figure 3.11). The glossiness of glazing is suggested by bright flecks of
paint; fabric folds by highlights and shadows; and, everywhere, the
mutual reflections of light between objects are indicated by subtle variations in colour. The wall is tinged with pinks and blues; the terracotta
vessel into which the maid pours milk is shaded with blue, suggesting an
interplay of light between it, the bobbled pitcher, and the tablecloth.
These variations in surface paint enhance the sense of depth in the scene.

The human visual system does, in fact, incorporate an understanding of the relationship between inter-reflections, three-dimensional structure, and spatial configuration, so that the perception of object colour and three-dimensional shape are intertwined, as the chromatic Mach card phenomenon demonstrates [36, 37]. The chromatic Mach card is a folded card, one side painted magenta, the other white (Figure 3.12).

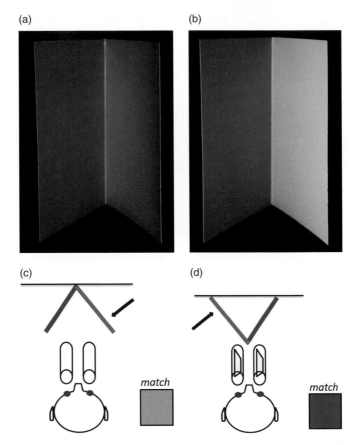

FIGURE 3.12 The chromatic Mach card. (a) and (b) Photographs of the same card, left side made of pink paper, right side of white paper, under the same light source, but from two different angles. In (a) there is less direct illumination on the white paper, so the indirect illumination it receives from the pink paper is more visible than it is in (b). (c) The card viewed through empty binoculars appears concave, and the indicated side, made from white paper, is matched to a pinkish white card. (d) The card viewed through Dove prisms appears depth-inverted and side-reversed, and the white paper side (now seen on the left) is matched to a deep pink. The insets show the appearance to a viewer.

When the card is seen as a corner – concave, in its veridical shape – the white side appears white, its pinkish glow unconsciously and accurately attributed to mutual reflections from the deeply coloured magenta side. When the apparent shape of the card is inverted so that it is now convex, with the folded edge protruding outwards towards the viewer, the white side now appears to be painted pink. (The depth inversion is achieved through use of a pseudoscope, binoculars fitted with Dove prisms which left–right reverse the image seen by each eye.) The visual system unconsciously infers that the pinkish glow cannot be attributed to mutual reflections, because, in its convex shape, the two sides of the card do not face each other; thus, the pinkness must be due to actual surface reflectance properties. In terms of a computational model, the visual system disentangles the components of illumination and surface reflectance in the reflected light signal by applying constraints of three-dimensional shape and making assumptions about the direction and spectrum of the light source [37].

Yet people differ in the matches they make to the white side of the card, some selecting very deep pinks, others pinkish whites. In contemporary vision science, the Bayesian probability framework has evolved from the 'inverse optics' approach and formalises the long-debated idea that seeing is subjective [38]. Like inverse optics, the Bayesian probability framework takes the stance that the visual system processes the image in order to reconstruct a three-dimensional world, but, unlike in inverse optics, there is no single uniquely accurate world, only a range of more or less probable alternatives. The Bayesian model of the chromatic Mach card accordingly predicts a range of possible colour matches to the white side of the card, depending on the internally assumed likelihood of the direction and spectrum of the light source, and the exact shape of the card, which are otherwise unknown to the viewer. Colour constancy must recruit similarly inferential processes. Because it is theoretically impossible to recover reflectance exactly, the visual system must make inferences. People operate as if colour provides a robust and reliable record of external reality, but it does not. Instead our brains infer properties from ambiguous images, actively constructing a plausible world, an internal reality that guides our behaviour in the physical world.

People therefore differ in the way they see colours because they can only guess at the underlying physical properties, it being impossible to access these exactly. Differences in individual experiences and constitution, as well as environmental conditions, will make their guesses differ.

Thus, the notion that colour *is* surface reflectance implies that people can never perfectly access the true colour of objects, but only approximate it, with greater or lesser accuracy. People – or, rather, their individual visual systems – may in that case be assessed on the extent to which the colours they see map onto invariant surface reflectance properties. Computational models of colour constancy may be assessed by the same measure. Indeed, for many purposes, the usefulness of the surface descriptor recovered by computer vision algorithms depends only on how close the descriptor is to surface reflectance. For automatic object recognition, for example in sorting apples, the descriptor should ideally be perfectly invariant under changing illumination, as it would be if it perfectly registered surface reflectance [39]. Yet, where the computational models purport to explain human vision, the surface descriptors should predict the colours that people perceive. In that case, the descriptors need not be perfectly invariant, because human colour constancy is not perfect, as decades of empirical measurements have proven. We will return to such laboratory measurements of colour constancy later.

Individual Variations in Colour Constancy: #thedress

The subjectivity of colour perception and individual variations in the way people resolve ambiguities inherent in images have nowhere been demonstrated more dramatically than in #thedress, a perceptual phenomenon that swept the Internet in spring 2015 [40] (Figure 3.13). A single overexposed photograph of a dress – with bolero jacket – generated worldwide disagreement over its colour: was the dress blue and black, or white and gold? 'I lost friends over that dress,' said a viewer on ITV's Gogglebox, watching the BBC's show *Colour: The Spectrum of Science*, for which my lab members and I performed a public experiment using the real dress [41]. The explosion of popular media coverage on #thedress

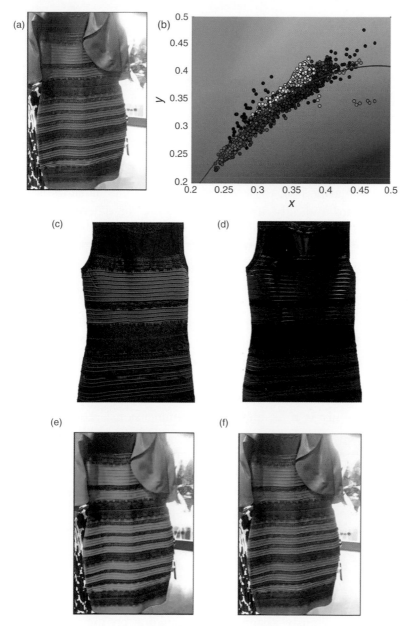

FIGURE 3.13 #thedress. (a) The original image. (b) The distribution of chromaticities in the original image, plotted in a CIE 1931 x–y chromaticity diagram. The blue line indicates the daylight locus. (c) and (d) Photographs of the real dress, illuminated by (c) two spectrally and spatially distinct light sources (blue and yellow) and (d) a single neutral source. (e) and (f) The original image rotated in the cone-opponent-contrast plane (Figure 3.14(b)) by (e) 270 and (f) 180 degrees, preserving the original pixel luminance values (assuming standard colour calibration). Photo (a) Cecilia Bleasdale. Photos (c) and (d) © Anya Hurlbert.

was rapidly followed by multiple scientific studies that quantified the differences in reported dress colour between people and tried to determine their causes [42–47]. Overall, these controlled laboratory and online studies showed that, when presented with a binary choice, the population was roughly evenly split between 'blue/black' (blue dress body, with black lace horizontal bands) and 'white/gold' (white body, gold lace), although the exact proportions varied. The phenomenon #thedress is unusual in the historical catalogue of colour 'illusions', because here the same image, viewed at the same time under the same conditions, evokes radically different perceptions in different people. Other colour 'illusions' show that relatively small changes in an image may evoke large changes in colour – such as a central square changing from grey to pink when its background changes from dark grey to green (see Figure 3.17(b) later in this chapter) – but these effects tend to be the same in all viewers.

The consensus amongst vision scientists is that people differ in the dress colours they report for the image because they differ in their inbuilt colour constancy processes: the underlying assumptions about the physical nature of the scene, the natural constraints their visual systems apply, and the image cues they unconsciously exploit [40]. The image is ambiguous: not only is it an imperfect two-dimensional rendering of a three-dimensional scene, with information lost due to overexposure and clipping, but also the illumination sources are hidden and their spectral properties uncertain, and the dress itself is an unfamiliar object made of unknown material (or at least it was at the time the photo appeared, although it is now worn by vision scientists the world over). Thus, there is no unique way to reconstruct the properties of the dress and the illumination which combined to make the photographic image. Different individuals may make different inferences about the light source illuminating the dress, for example, and these will entail different conclusions about the dress material itself. In fact, as several studies have shown, people who unconsciously assume the illumination on the dress to be bluer and darker tend to see the dress as white/gold, whereas those who see the illumination as yellower and brighter see the dress as blue/black [42, 46–48]. Both combinations in reality would give rise to the observed image, in which the pixel values of the dress fall on a bluish–yellowish line in the chromaticity plane[1] (Figure 3.13(b)). This particular distribution of chromaticities is not incidental to #thedress, but in fact a key driver of the phenomenon. We will return to this observation later.

The phenomenon exemplified by #thedress is not, tellingly, confined to the photographic image alone. The ambiguity may be reproduced in a real scene, thus demonstrating that the human visual system operates in a continual inferential mode in everyday behaviour, not just when interpreting flat pictures. For the BBC show [41], and in other public experiments, we illuminated the real dress (a viscose-polyamide lace detail bodycon dress by Roman Original, in its 'Royal Blue' colourway) with two spatially and spectrally separated light sources, generated by spectrally tuneable multi-channel LED lamps: one a diffuse bluish light, the other a more focused yellowish light (Figure 3.13(c)). Different individuals, viewing the dress at the same time, gave it different colour names, many calling it blue/black, but others calling it white/gold, or, when allowed to choose any combination, reporting it as blue/brown, violet/gold, violet/black, or other colour names. Even though it was obvious that there were multiple light sources in the scene, these colour names and their variability suggested that people's visual systems unconsciously assumed a single, broadband light source, but differed in their assessment of its spectrum.

The vehemence with which people clung to their dress colour names shows how deeply people feel that colour is an integral, invariant part of objects. If colour is a physical property, it should not be a matter of opinion, but one of fact. Hence, people were forced to question either the personal integrity of their friends or their own faith in visual perception as a passive recorder of factual reality. Yet, as the brilliant but irascible Ewald Hering so presciently observed, people would not be able to develop such faith in colour perception without the internal workings

---

[1] I use the term chromaticity to describe a physical property of the light spectrum, rather than a human perceptual response. The chromaticity is obtained from the integral over wavelength of the product of the spectral power distribution of the light with the colour matching functions of the human visual system, which are themselves linear combinations of the spectral sensitivity functions of the retinal cones in an average trichromatic observer (with an additional normalisation step). It is therefore a quantitative description of the light spectrum, but does not uniquely specify the complete spectral power distribution.

of colour constancy: 'If the colours of objects ... continuously changed ... along with the illumination changes, then it could not happen that individual objects have fixed colours for us which we regard as essential properties of the objects and which we call their real colours' [49]. And colour constancy requires active perceptual processes, not merely passive transmission of the images formed at the eye.

The Fundamentals of Colour Vision

The individual variability of colour perception arises not only from differences in the processes of colour constancy, but also from differences in the basic sensory machinery of colour vision, beginning with the optical equipment of the eye. Human colour vision is normally trichromatic, meaning that people possess three distinct cone photoreceptor types which operate at daytime light levels, each responsive to a different but overlapping broadband region of the spectrum (the S, M, and L cone types, sensitive to short-, middle- and long-wavelength bands, respectively). This trichromatic visual system evolved from a primitive dichromatic system, whose two cone types enabled discrimination between short- and long-wavelength spectral bands only (Figure 3.14) [50, 51]. About 30 million years ago, the gene encoding the long-wavelength receptoral pigment mutated into two, giving rise to our 'modern' M and L cone types.

Two hypotheses suggest different drivers for the eruption of these two cone types: to enable our primate ancestors to discriminate reddish–yellowish, ripe, palatable fruits and leaves against a less edible green background [52–54]; or – given that around the time of this mutation our ancestors became bare-skinned – to enable our ancestors to distinguish subtle changes in blood oxygen level of the face, in turn enabling them to discern the health or emotional state of other individuals from skin colour changes [55]. Possibly the most plausible scenario is that initially frugivory drove the mutation, and our ancestors then capitalised on the ability conferred by trichromacy to read social–sexual signals conveyed by skin colour [56].

The ancient dichromatic system itself might have been designed not for seeing in the sense that we know it now, enabling conscious

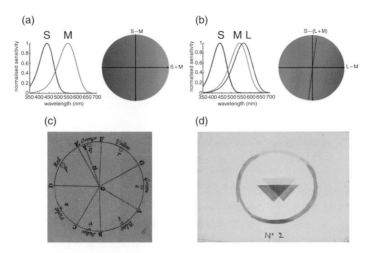

FIGURE 3.14 Spectral sensitivities of cone photoreceptors in (a) 'ancient' dichromacy and (b) 'modern' trichromacy. For each set of spectral sensitivities, the adjacent diagram plots the chromaticities corresponding to the coordinates of cone-opponent channel activation (the cone-opponent contrast plane) (assuming standard colourimetric calibration), with the daylight locus (blue line). (c) Newton's colour circle. (d) Joseph Mallord William Turner (*c.* 1842–1848), from *II. Various Perspective Diagrams, Lecture Diagram: Colour Circle No.2.* Graphite and watercolour on paper. Tate, London. Photo © Tate.

identification and exploration of external objects, but for initiating reflexive behavioural responses to changes in light spectra [57]. For our primitive land-dwelling ancestors, the spectral changes in illumination at dawn and dusk would be critical. Monitoring these would help to entrain the body's rhythms to the external light–dark cycle [58].

The notion that the ancient dichromatic system served to sense changes in the ambient illumination spectra is supported by the close alignment between the physiology of the former and the physics of the latter. Daylight is a mixture of sunlight and skylight, varying with time of day, season, and geographical location, and its varying chromaticities fall along a curve close to the Planckian or black-body radiation locus, from bluish to orangish yellow. The daylight locus in turn lies close to the line of chromaticities that exclusively modulate the short-wavelength cone activity, the S-cone isolating or tritanopic line (Figure 3.14). Changes from one chromaticity to another along this line alter the activity in the S cones only, maintaining constant activity in the L (and M) cones. This proximity suggests that neurons receiving S-cone input might be specialised for following changes in the daylight spectrum. Indeed, other evidence suggests that cortical neuronal sensitivities are skewed more towards the daylight locus than the pure S-cone isolating axis [59].

A fundamental process in colour vision is cone-opponency, the comparison of light signals received by the distinct cone types. Because the responses of the L and M cone types are highly correlated – their peak sensitivities differ by only about 30 nm [60] – the important information for discriminating between spectral signals lies in the difference between the cone responses, rather than in their absolute values [61]. Broadly speaking, the responses of neurons that register the difference between the L and M cone responses – $L - M$ cone-opponent neurons – therefore are well-suited to signal differences between surface properties of objects, whereas neurons that register relative changes in the S cone responses – the S vs. $(L + M)$ cone-opponent neurons – are suited for signalling illumination changes. Indeed, further analysis of object properties supports this idea.

Natural objects are not perfectly uniform in their surface properties: the heterogeneity of pigment particles in the skin of an apple (e.g. chlorophyll and carotenoids) makes its reflected light spectrum differ at different locations across its surface, and the curvature of the apple's surface also means that the total amount of light it receives and reflects will vary with position. Thus, the apple presents to the eye not a single flat block of unvarying light, but a shaded, textured, contoured light pattern. Yet, to the human visual system, the collection of light signals from the apple's surface forms a characteristic signature, a linear cluster, in the chromaticity plane formed by the two cone-opponent dimensions, $S - (L + M)$ and $L - M$ (Figure 3.15) [62, 63]. Most natural objects form similar signatures, and mostly in the lower right quadrant of the cone-opponent plane, thereby defining 'warm' colours [64]. Further, these signatures remain roughly constant under changes in illumination, because, in the normalisation and decorrelation of the L and M responses, the contributions of the illumination are factored out [63] (Figure 3.15). Colour constancy is thereby born from cone-opponency. Indeed, the

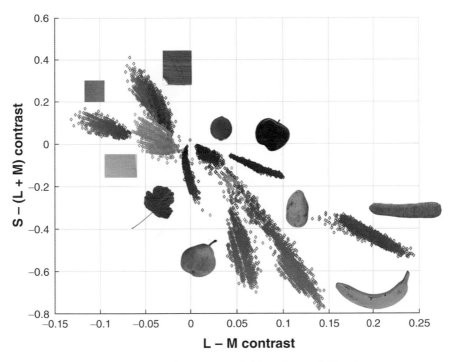

FIGURE 3.15 The distribution of object chromaticities, in cone-opponent contrast space. Objects from right to left: carrot, banana, potato, apple, lime, pear, cloth 1, leaf, cloth 2, cloth 3.

stable angle of the signature, readily computable from the ratio of the $L - M$ and $S - (L + M)$ responses, might be the neural correlate of object colours [63]. Thus, there seems to have been an evolution from crude vision for sensing changes in light, on basis of the opponency between the primordial S and L cone types, to fine vision for finding objects and discerning their material properties, on the basis of modern $L - M$ opponency.

The two cone-opponent directions, $S - (L + M)$ and $L - M$, have traditionally been called 'blue–yellow' and 'red–green', in tacit support of Hering's theory that colour perception is founded on the existence of four unique sensations corresponding to red, green, blue, and yellow. Yet the cone-opponent axes do not align with those connecting the unique hues [65], and the evidence for physiological mechanisms which do underlie perceptually unique hues has remained elusive. Instead, the cone-opponent axes connect chromaticities which would more aptly be named 'lime–lavender' $(S - (L + M))$ and 'red–cyan' $(L - M)$ [66].

Long before vision scientists began to tackle these complications, the perceptual opponencies inherent in the neural processing of colour were observed and plumbed by painters for all the vigour and resonance they would lend visual art. Artists were guided by colour circles, after the first proposed by Newton, in his *Opticks* of 1704, the first systematic arrangement of colours into a circle where neighbours are meaningfully related, and which embeds the idea of complementarity (Figure 3.14(c)). In Newton's circle, red opposed blue, yellow opposed violet, and green opposed purple. Moses Harris, in his *Natural System of Colours*, proposed a different opposition, maintaining the primary positions of red, blue, and yellow, yet pitting them against green, orange, and purple, respectively. Goethe posited a similar opponency, but was obsessed with the polarity between bright and dark. That obsession, and his observations of afterimages, the illusory images elicited by a blank canvas following prolonged viewing of an image, appearing in colours complementary to the original, influenced Hering's opponent-colour theory. (See also Figure 3.19.)

The fact is that scientific definitions of which colours complement or oppose each other, as well as the method of how to define them – for example, as pairs of colours that sum to white, or pairs which, when spatially juxtaposed, most heighten dissimilarity between them, or as afterimage couplings – vary almost as much as do painters' personal definitions and deployments. Chevreul, a chemist called into the Gobelin's dye factory to explain why yarn colours appeared to change when woven in with other yarn colours, observed in 1868 that '[i]n the case where the eye sees at the same time two contiguous colours, they will appear as dissimilar as possible, both in their optical composition and in the height of their tone' [67]. This jarring dissimilarity is expressly what van Gogh wished to achieve with *The Night Café*, 1888, brashly juxtaposing intense, saturated greens and reds against a restless yellow ground, which he described as 'one of the ugliest I have done', 'express [ing] the terrible passions of humanity by means of red and green' [68] (Figure 3.16).

FIGURE 3.16 Vincent Van Gogh. *Le café de nuit* (*The Night Café*), 1888. Oil on canvas. Photo: Yale University Art Gallery.

Chevreul's observations, which he formalised in his book *The Laws of Contrast of Colour* [67] (Figure 3.17(a)), express the second fundamental opponency in colour perception: not only are the initial cone responses at the same retinal locations converted into relative activations through the second-stage cone-opponent comparisons, but also these relative responses are compared against each other across space. The processes of spatial chromatic contrast encode differences in cone-opponent responses from neighbouring parts of the image, thus, for example, inducing pinkness into a grey square placed on a green background (Figure 3.17(b)). Spatial scale is critical for chromatic contrast: on a coarse scale, a block of surrounding colour will induce its opposite in an enclosed patch (Figure 3.17(b)), whereas two colours, interleaved on a fine spatial scale, will assimilate: the grey checkerboard adopts the green of its counterpoint (Figure 3.17(c)).

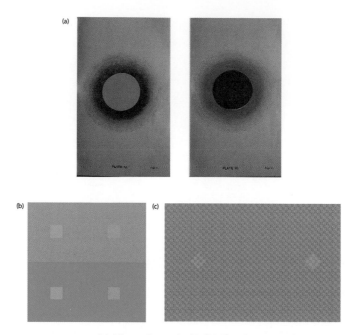

FIGURE 3.17 (a) Plates from [67]. (b) Simultaneous chromatic contrast. The pairs of small squares in the upper and lower halves are identical, yet the right-hand square appears grey against the dark grey background and pink against the green background, and the left-hand square appears green above and gold below. (c) Assimilation. The nine small grey squares on the left and right (in diamond formation) are identical, but appear tinged pink and green, respectively. Photos © Anya Hurlbert.

The pointillists, it might be said, emphasised assimilation and complementary additivity, whereas expressionists such as Mondrian, van Doesburg, and Klee concerned themselves with contrast. This dependence of colour on its spatial context is another reason for its dismissal by the Rubenistes and others on the *disegno* side of the divide; colour's susceptibility to its surround made it less trustworthy and meaningful than form. In his mature works, though, Mondrian's aim was to destroy this contextual dependence and achieve 'mutual equivalence' between colours and forms, laying down 'pure and determinate' blocks of colour divided by straight black lines [69] (Figure 3.18).

Seemingly paradoxically, the mechanisms that make colour depend on its spatial context are the same ones that contribute to colour

(a)   (b)

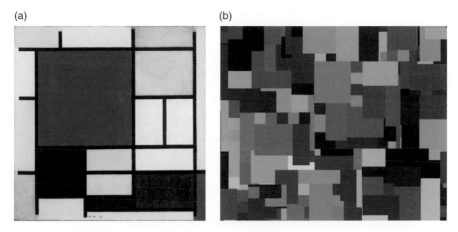

FIGURE 3.18 (a) Piet Mondrian. *Composition with Large Red Plane, Yellow, Black, Grey and Blue*, 1921. Oil on canvas. Collection of the Gemeentemuseum Den Haag. © 2020 Mondrian/Holtzman Trust. (b) Typical 'Mondrian' stimulus used in colour constancy experiments.

constancy [70, 71]. Colour constancy is a property of surfaces seen in the context of other surfaces only. A solitary uniform patch in an otherwise black environment – 'in the void' – will not remain constant in appearance if the light spectrum it reflects alters. Colour constancy relies on comparisons between light signals emanating from different surfaces, or parts of surfaces, in the scene. Spatial chromatic contrast instantiates one such method of comparison. If the background reflectances average to grey, then subtracting out the background chromaticity would factor out the illumination chromaticity. In the example shown in Figure 3.17(b), it is as if the visual system treats the green background as a grey surface illuminated by green light, and therefore interprets the grey square as a pink reflectance under the same light. The 'grey-world' assumption is an example of a natural constraint the visual system might apply in resolving the inherent ambiguity in reflected light. It is also a critical feature of many computational models of colour constancy that rely on space-averaging to estimate the illumination spectrum [72].

Colour also depends on temporal context, in that the sensitivities of the cones continuously adjust according to the level of their stimulation,

FIGURE 3.19 The Dunstanborough Castle illusion. Upon fixating on the central dot of the upper-left pseudocoloured version of an original photograph of Dunstanborough Castle, Northumberland, for about 15 seconds, and then transferring one's gaze to the central dot in the upper-right image, a luminance-only version of the photograph, yields an afterimage with the veridical colours of the original. The illusion works best if each image is viewed centrally, and fills the viewing field. Lower panel: The original photograph, courtesy of Richard Gregory. Upper panel: Photos © Anya Hurlbert.

as do the neurons to which they connect. This process of chromatic adaptation helps to neutralise responses to chromatic biases in the illumination. For example, in reddish light, the response gain of the long-wavelength cones will be proportionately reduced, cancelling out the effects of the additional reddish light reflected from surfaces. Chromatic adaptation upsets the balance of cone-opponent mechanisms, which explains why coloured afterimages occur in response to a subsequent achromatic stimulus (Figure 3.19). The perceptual effects of chromatic adaptation thereby may also reveal the neural characteristics of cone-opponency.

Chromatic contrast and adaptation operate in early stages of visual processing, via adaptive scaling or lateral interactions of receptors and low-level neurons in response to point-by-point image signals [73, 74], without the need for more meaningful image interpretation or multi-layered neural processing [72, 75]. These are powerful contributors to colour constancy, hard-wired in at low levels, and essentially impervious to conscious manipulation [76, 77]. But they are not the only mechanisms at play [78]. On other levels, image features such as glossy highlights, or mutual reflections, may assist in estimating the illumination [72], even as colour information helps to identify these features. At the cognitive level, stored knowledge about the typical appearance of familiar objects may also influence colour constancy not only of those objects directly but also of unfamiliar objects nearby. Variations in the familiar colour of bananas or human skin, for example, may guide the visual system to variations in illumination colour [63, 79]. Monet's depictions of illumination in his series paintings thus signify an astonishing penetration into pre-conscious perception, reaching beneath these many layers of contributory mechanisms to colour constancy.

It was Hering, again, who pinpointed the cognitive contributions to colour constancy. He defined the memory colour of a familiar object as '[t]he colour in which we have most consistently seen an external object', which 'becomes a fixed property of the memory image' and is then 'a further influence on the way we see [via] "psychological" factors ... which depend on individual experiences already established in the nervous substance' [49]. In everyday life, memory colours may provide subtle but effective support to the strong forces of adaptation and contrast [80]. In the Dunstanborough Castle illusion (Figure 3.19), they dramatically boost the effects of chromatic adaptation.

Dunstanborough Castle was another of Turner's favoured sites in the north of England, whose visage he reworked in 1798 from earlier sketches, freely modifying his own memory for dramatic effect: the stormy sea fantastically floods the foreground, isolating the ruined fortress and emphasising its bleakness (Figure 3.20).

Turner, taking up Goethe's emphasis on light–dark polarity, also devised his own colour circle, which was not a hue circle proper but a pyramid, in which yellow reigned supreme at the top (Figure 3.14(d)). He

FIGURE 3.20 J. M. W. Turner. *Dunstanborough Castle, c.* 1798. Oil on canvas. Collection of the Dunedin Public Art Gallery.

flattened Harris's circle: 'Sink the yellow until it light into the red and blue, and hence two only: light and shadow, day and night, or gradation of light and dark' [3]. In homage to Goethe, Turner created late in life his famous pairing of paintings *The Morning after the Deluge*, a cyclone of brilliant colours, converging on yellow and white in the centre, and *The Evening of the Deluge*, blackness encircling a grey–blue core (Figure 3.21). Turner's genius was to conjure up the multiple modalities that colour possesses. With pigment alone, limited by subtractive mixing, Turner brought forth colour not only as a surface attribute, evoking the material properties of objects, but also as an extended property of voids, volumes, and lights, thus presaging the abstract movement in art.

Measuring Colour Constancy

Turner's *Morning* and *Evening* paintings also express the concept of time, through the colour of light. As the sun changes in elevation throughout the day, so the spectrum of daylight varies, as noted earlier. Turner

(a) (b)

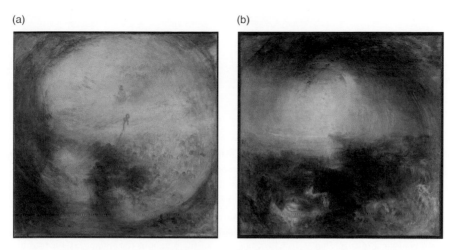

FIGURE 3.21 J. M. W. Turner. (a) *Light and Colour (Goethe's Theory) – the Morning after the Deluge*, 1843. (b) *Shade and Darkness – The Evening of the Deluge*, 1843. Oil on canvas. Tate, London. Photos © Tate.

captured not only the variation in daylight in his paintings, almost explicitly laying down the daylight locus in his *Colour Beginning* (1819) (Figure 3.22), but also its potency in conveying non-visual phenomena such as time of day or season [81].

Because the spectral characteristics of daylight are regular and predictable, and daylight would have been the sole source of illumination for the millions of years during which colour vision was honed, it is plausible to assume that colour constancy mechanisms were optimised during evolution for illumination changes along the daylight locus. To assess the validity of this assumption, we need a way to assess the integrity of colour constancy itself. Just as there are multiple neural mechanisms that have been proposed to underlie colour constancy, so are there many methods for empirically measuring the perceptual phenomenon. Although their results differ in detail, the many distinct studies agree in principle: colour constancy is not an all-or-none phenomenon. Some surfaces remain more stable in colour under some illumination changes than others, and perfect colour constancy is rare.

A typical method used to measure colour constancy in the laboratory is asymmetric matching. People are asked to view two images, separated in

(a)

(b)

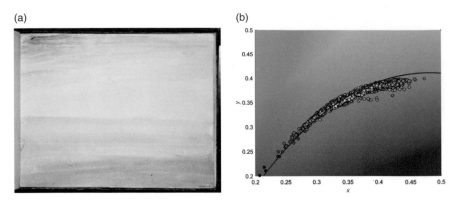

FIGURE 3.22 (a) J. M. W. Turner. *Colour Beginning, from Como and Venice Sketchbook [Finberg CLXXXI], A Beach and the Sea below a Pale Cloudy Sky,* 1819. Watercolour. Photo © Tate. (b) Chromaticities of the pixels in the image (a), plotted in the CIE 1931 *x–y* chromaticity plane (assuming standard calibration). They fall near the daylight locus (solid blue curve).

space and/or time, each of which depicts a scene under a controlled illumination. Edwin Land and John McCann invented 'Mondrian' stimuli, two-dimensional arrays of irregularly sized rectangles, each uniform in chromaticity and luminance, but different from its neighbours [82] (Figure 3.18(b)). (These images, with their varied distribution of chromaticities, complete coverage of the ground, and lack of borders between neighbouring surfaces, resemble Mondrian's earlier works more than his iconic Neo-Plastic primary-colour compositions.)

In a typical constancy experiment of the late twentieth century [83] (for a review see [84]) two Mondrian images are displayed on a computer screen. One Mondrian serves as the reference; its surfaces are simulated under a reference illumination, for example, neutral daylight (Figure 3.23). (That is, the RGB colours are calculated to elicit the same set of receptoral responses as would a surface with a particular spectral reflectance function illuminated by a light with equal power at each wavelength. This is the beauty of the trichromatic visual system.) In the other Mondrian, the test image, the same set of surface reflectances, with the same spatial configuration, are simulated under a test illumination. In the reference image, one surface colour is left unspecified, and

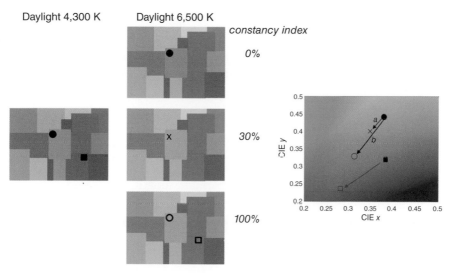

FIGURE 3.23 Colour constancy index calculation from results of the asymmetric colour matching paradigm. Left-hand panel: a Mondrian scene illuminated by daylight of correlated colour temperature (CCT) 4,300 kelvins (4,300 K). Central panel: a Mondrian scene consisting of same surface configuration of surface reflectances as on the left, showing three example colour matches made by a participant to the central patch, with associated colour constancy indices. Right-hand panel: chromaticities of example matches (x,o) and of another surface (marked by corresponding symbols in the patch corner). Filled symbols: chromaticity under 4,300 K; empty symbols: chromaticity under 6,500 K. The (simplified) colour constancy index is calculated as a/b, where the distances are measured in a perceptually uniform colour space. (The CIE 1931 x–y chromaticity plane shown here is not perceptually uniform.) In a typical experiment, only one scene under each illumination would be presented to the participant, without symbols, either simultaneously, separated in space, or successively.

the participant is asked to adjust its colour to match the colour of the corresponding surface in the test image.

If the participant were perfectly colour constant, she or he would select the RGB value predicted by the simulation, that is, the RGB value calculated from multiplying the matching surface reflectance by the reference illumination power spectrum. With no colour constancy, the participant would select an RGB value identical to that in the test image, that is, taking no account of the change in illumination. The difference between the participant's RGB match and the perfect constancy match indicates the deviation from perfect colour constancy. When normalised by the distance between the no-constancy match and the perfect constancy match, the extent of colour constancy may be expressed as a percentage (Figure 3.23). Over the past four decades, from typical methods such as these, deployed in a variety of empirical studies, colour constancy has been measured to be anywhere from 10 to 85 per cent [84].

A related method for measuring colour constancy is achromatic adjustment. Instead of adjusting chromatic surfaces to match their partners in a differently illuminated scenes, the participant adjusts the chromaticity of only one surface, to make it appear white. The chromaticity that the participant selects – his or her achromatic point – then indicates the chromaticity of the illumination he or she perceives. The rationale is that a perfectly white surface will take on the chromaticity of the ambient illumination because it reflects all of the illumination incident on it at every wavelength. Because of colour constancy, a 100 per cent reflective white surface continues to appear white despite reflecting bluish light under bluish ambient illumination, and yellowish light under yellowish ambient illumination. As Renoir instructed, 'The blue of the sky must show up in the snow' [85]. In Monet's *Lavacourt under Snow* (Figure 3.24), the snow appears white, although there is no part of the painting that is pure white, instead only the blues, violets, and yellows of daylight and shadow, with myriad other reflections.

Therefore to evince perfect colour constancy, a person should perceive as white a surface that reflects exactly the chromaticity of the ambient illumination. The achromatic point thus may be used as a measure of colour constancy: the closer the achromatic setting to the actual chromaticity of the ambient illumination, the greater the colour constancy.

Measured in this way, colour constancy is still not perfect, but, again, the deviation from constancy depends on the illumination, the person, and the particulars of the paradigm and stimuli. There is large variability within and between people in the chromaticity they select as white, yet there is consistency in the variability: the spread tends to be greatest along a blue–yellow direction paralleling the daylight locus [86, 87]. Furthermore, under neutral illuminations, or 'in the void', people on average tend to adjust their white points to be slightly blue. Similarly, when presented with small patches of light on a darkened computer screen, people tend to name desaturated blues as white but desaturated

FIGURE 3.24 Claude Monet. *Lavacourt under Snow*, 1878–1881. Oil on canvas. © The National Gallery, London. Sir Hugh Lane Bequest, 1917.

yellows as yellow [88]. Bluish whites tend to be seen as whiter than warmer whites, at least in Western cultures. Under more chromatic illuminations, whether simulated on a computer screen [89] or experienced in a real setting [90, 91], people's whitepoints show larger deviations from the illumination chromaticity, as if colour constancy has been pushed beyond its limits. Yet, again, the whitepoints tend to deviate in the same direction, towards the daylight locus [89, 91].

Colour constancy is therefore not unequivocally 100 per cent. Yet it is also clear that colour constancy is not the same for all surfaces under all illumination changes, and that changes along the daylight locus may receive special treatment. A plausible hypothesis is that colour constancy is best for those illuminations under which the trichromatic visual system of humans evolved: natural daylight. To test this hypothesis requires comparing colour constancy under changes in illumination along the daylight locus with that under other directions of illumination change. One method for doing so is the global illumination discrimination task (IDT) [92]. In this task, we assess how much the illumination may be changed between two views of a scene without the participant detecting any change at all in the scene (Figure 3.25). Perfect colour constancy would entail perfect scene stability, and no awareness of the illumination change.

This paradigm is made possible by recent advances in lighting technology. The real light sources we use are spectrally tuneable multichannel LED luminaires, whose output light may be sculpted smoothly in real time by altering the relative contributions of the different LED emissive channels, producing an almost infinite set of spectra (Figure 3.25). Changes of any size in the illumination spectrum may therefore be generated, from very small to large. For the standard IDT, we compared changes in illumination chromaticity along the daylight locus, from bluish to yellowish, with changes in the orthogonal direction, from reddish to greenish, to determine the largest magnitude of change in each direction which remains undetectable. Conversely, this magnitude therefore also defines the just noticeable change in illumination, the threshold.

Critically, we find that illumination-change discrimination thresholds are not the same for every direction [92, 93]. Thresholds tend to be largest in the bluish direction. That is, people find it harder to see changes towards bluer illumination chromaticities, or, conversely, the scene appears more stable as the illumination changes towards blue. Thresholds are smallest along the orthogonal reddish–greenish direction; people find it easiest to see illumination changes towards red or green. This finding is true not only for experiments performed in the laboratory light-room with small numbers of participants, each one performing hundreds of trials, but also in a real-world setting with very large numbers of participants, each performing only two trials.

As part of the National Gallery's 2014 summer exhibition Making Colour, we ran a public experiment in its cinema. We displayed a pair of artworks on the central front wall, both reproductions of Paul Gauguin's *Bowl of Fruit and Tankard before a Window* (NG6609), one hand-painted in oils by Gabriella Marcaro and the other a 12-ink print of Marcaro's

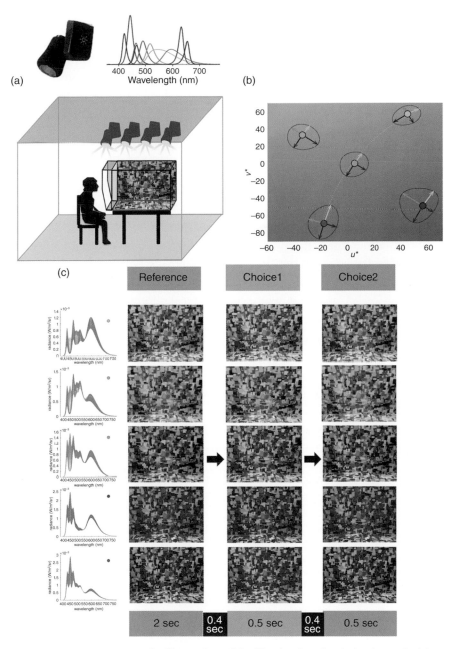

FIGURE 3.25 An illustration of the illumination discrimination task. (a)
The participant sits in a white-painted light-room, viewing a Mondrian-papered scene,

rendition. Both were illuminated by a bank of spectrally tuneable multi-channel LED luminaires, the sole source of light for the entire cinema apart from the two video screens which played an interactive film about colour perception every 15 minutes throughout the day. We interleaved a version of the IDT with the film, asking audience members to provide responses using individual wireless clickers, and, over the course of the three-month exhibition, ran 1,440 sessions. In different sessions, we tested a different type and amount of illumination change, and then collapsed the results across all participants, who numbered more than 10,000. Just as for the small groups of about 10 participants viewing manufactured Mondrians or grey walls in controlled laboratory experiments, these participants, viewing a post-Impressionist still-life painting, found it hardest to distinguish bluish changes in illumination.

The results suggest that colour constancy is better for illumination changes that mimic the natural changes of daylight, in the bluish and yellowish directions, and worse for atypical illumination changes that would less frequently have been encountered in human evolution. Further experiments using the IDT paradigm reveal nuances to this simple statement. The precise pattern of illumination discrimination thresholds depends both on the spectral properties of the reference illumination and on the surfaces in the scene. For on-average neutral scenes under chromatic illuminations – red, yellow, green, or blue – illumination discrimination thresholds tend to be largest in the direction away from the illumination chromaticity [94] (Figure 3.25(c)). In general, changes in illumination chromaticity towards neutral are hardest to discriminate. Yet, over all illumination changes, there is still a 'blue bias': on average,

CAPTION FOR FIGURE 3.25 (cont.) lit by spectrally tuneable multi-channel LED lamps (drawing not to scale). (b) In each trial (each row), the scene is lit first by a reference illumination, and then, after a dark interval, successively by two alternative illuminations, also separated by a dark interval, one of which is shifted along one of four chromatic directions from the reference. The leftmost column shows spectra of alternative illuminations along the blue and yellow directions. The participant selects which alternative matches the reference illumination. (c) The size of the arrow indicates the just noticeable difference (discrimination threshold) from the reference illumination chromaticity (central disc) in each chromatic direction (in colour space CIELUV; thresholds multiplied by 1.5 for visibility).

changes towards bluish daylight are least discriminable. Given the need for the human visual system to adopt helpful prior assumptions about environmental conditions in order to resolve image ambiguities, it seems likely that this 'blue bias' results from an unconscious internal assumption that illuminations are often bluish daylights. When tasked to remember previously encountered illuminations, this internal bias may cause the memory to decay towards blue.

To return to #thedress, the fact that the chromaticities of the image pixels fall along the daylight locus is therefore significant, and indeed is crucial in eliciting the dramatic differences between people's perceptions of its material constitution. Flipping the blues to greens and the yellows to reds to create a new green/red image of the dress destroys its polymorphism [44, 88] (Figure 3.13(e)). The phenomenon depends on the particular chromaticity distribution of the original image. It seems that, because the neural mechanisms underlying colour constancy are optimised for illuminations along the daylight locus, the spread of daylight chromaticities in the image increases the uncertainty over the particular illumination present. Bluish illuminations are more difficult to discriminate from white [92, 94], and therefore blues, in particular, are more likely to be seen as white surfaces under bluish daylight [88]. Thus, 'white/gold' is an entirely plausible interpretation of #thedress. 'Blue/black' is also plausible, given the likelihood of yellowish daylights. Yet, swapping the yellow and blue chromaticities while preserving their luminances, so that the dress body becomes light yellow and the lace dark blue (Figure 3.13(f)), largely eliminates the perceptual discrepancy between individuals, almost all now agreeing that the body is yellow or gold. One explanation for this effect is that bright, warm colours are the preserve of objects rather than lights, but #thedress phenomenon still needs further exploration.

The blue bias in illumination perception and colour constancy is mirrored by an emotional bias for blue. Individual preferences for hues tend to peak in the blues, across cultures and ages [20]. Do people love blue because it is the colour of light? Turner loved as a boy to lie watching the sky: 'I used to lie for hours on my back watching the skies, and then go home and paint them, and there was a stall in Soho Bazaar where they sold drawing materials and they used to *buy* my skies' [95].

The emotional associations Turner made with the colour of sky will have been profound: supine, relaxed, he drank in the deep blue, entranced by the vicissitudes of the clouds, and then turned his pleasure into profit.

Turner's red buoy in *Helvoetsluys* thus carried meaning on multiple levels. It delivered chromatic contrast, in a grey–blue sea. Yet, in his opposition to colour circles founded on hue complementarity, Turner refused to acknowledge the conventional contrasts of red against green, blue against orange, yellow against violet. Instead, in Turner's own colour diagram, according to John Gage, yellow 'functioned as a light, and red and blues as darks', although he remained 'ambivalent' about red's luminosity, instructing an engraver to 'sometimes translate [red] into black, and another time into white. If a bit of black gives the emphasis, so does red in my picture' [1]. His resistance to opponent-hue schemes might in fact have arisen from his love of light, and, as we have seen, nowhere is daylight's fundamental blue–yellow opponency more evident than in Turner's masterful renderings of light itself. The red mark in *Helvoetsluys* is a blunt object intruding on a turbulent sea of light; it was the weapon of an intellect honed on the emotion of colour. As Constable said, with that red mark, 'He has been here, and fired a gun' [1].

References

[1] Gage, J., *J. M. W. Turner: 'A Wonderful Range of Mind'*. New Haven, CT and London: Yale University Press, 1987.

[2] Hamilton, J., *Turner's Britain*. London: Merrell Publishers Limited, 2003.

[3] Gage, J., *Colour and Culture*. London: Thames and Hudson, 1993.

[4] Ball, P., *Bright Earth: The Invention of Colour*. London: Viking, 2001.

[5] Biederman, I., and Bar, M. One-shot viewpoint invariance in matching novel objects. *Vision Res.* 1999; 39(17): 2885–2899.

[6] Biederman, I. Recognition-by-components: A theory of human image understanding. *Psychol. Rev.* 1987; 94(2): 115–147.

[7] Poggio, T., and Edelman, S. A network that learns to recognize 3-dimensional objects. *Nature* 1990; 343(6255): 263–266.

[8] Biederman, I., and Ju, G. Surface versus edge-based determinants of visual recognition. *Cogn. Psychol.* 1988; 20(1): 38–64.

[9] Tanaka, J. W., and Presnell, L. M. Color diagnosticity in object recognition. *Perception & Psychophys.* 1999; 61(6): 1140–1153.

[10] Oliva, A., and Schyns, P. G. Diagnostic colors mediate scene recognition. *Cogn. Psychol.* 2000; 41(2): 176–210.

[11] Therriault, D. J., Yaxley, R. H., and Zwaan, R. A. The role of color diagnosticity in object recognition and representation. *Cogn. Processing* 2009; 10(4): 335–342.

[12] Rossion, B., and Pourtois, G. Revisiting Snodgrass and Vanderwart's object pictorial set: The role of surface detail in basic-level object recognition. *Perception* 2004; 33(2): 217–236.

[13] Naor-Raz, G., Tarr, M. J., and Kersten, D. Is color an intrinsic property of object representation? *Perception* 2003; 32(6): 667–680.

[14] Hurlbert, A., and Poggio, T. A network for image segmentation using color. In Touretzky, D. S., ed. *Neural Information Processing Systems I.* San Francisco, CA: Morgan Kaufmann, 1989; 297–303.

[15] Nieuwenhuis, C., and Cremers, D. Spatially varying color distributions for interactive multilabel segmentation. *IEEE Trans. Pattern Analysis Machine Intell.* 2013; 35(5): 1234–1247.

[16] Kingdom, F. A. A., Perceiving light versus material. *Vision Res.* 2008; 48(20): 2090–2105.

[17] Wolfe, J. M., and Horowitz, T. S. What attributes guide the deployment of visual attention and how do they do it? *Nature Rev. Neurosci.* 2004; 5(6): 495–501.

[18] Jung, C. G. *Archetypes and the Collective Unconscious.* Vol. 9 of *The Collected Works of C. G. Jung*, ed. Fordham, M., Adler, G., and Read, S. H. Princeton, NJ: Princeton University Press, 1959.

[19] Humphrey, N. K. The colour currency of Nature. In Porter, T., and Mikellides, B., eds. *Colour for Architecture.* London: Studio-Vista, 1976; 95–98.

[20] Hurlbert, A., and Owen, A. Biological, cultural, and developmental influences on color preference. In Elliot, A. J., Fairchild, M. D., and Franklin, A., eds. *Handbook of Color Psychology.* Cambridge: Cambridge University Press, 2015; 454–480.

[21] Goethe, J. W. von. *Theory of Colours.* London: John Murray, 1987.

[22] Critchley, M. Acquired anomalies of colour perception of central origin. *Brain* 1965; 88: 711–724.

[23] Livingstone, M. S., and Hubel, D. H. Anatomy and physiology of a color system in the primate visual cortex. *J. Neurosci.* 1984; 4(1): 309–356.

[24] Shipp, S., Adams, D. L., Moutoussis, K., and Zeki, S. Feature binding in the feedback layers of area V2. *Cereb. Cortex* 2009; 19(10): 2230–2239.

[25] Seymour, K., Clifford, C. W. G., Logothetis, N. K., and Bartels, A. Coding and binding of color and form in visual cortex. *Cereb. Cortex* 2010; 20(8): 1946–1954.

[26] Lafer-Sousa, R., and Conway, B. R. Parallel, multi-stage processing of colors, faces and shapes in macaque inferior temporal cortex. *Nature Neurosci.* 2013; 16(12): 1870–1878.

[27] Chang, L., Bao, P. L., and Tsao, D. Y. The representation of colored objects in macaque color patches. *Nature Commun.* 2017; **8**: 2064.

[28] Schloss, K. B., Lessard, L., Walmsley, C. S, and Foley, K. Color inference in visual communication: the meaning of colors in recycling. *Cogn. Res.: Principles Implications* 2018; **3**: 5.

[29] BBC, *Witness: Bloods and Crips.* 2015.

[30] Hinton, J. *Lilac Chaser.* 2005; available from https://michaelbach.de/ot/col-lilacChaser/

[31] Hurlbert, A. C. The perceptual quality of color. In Albertazzi, L. ed. *Handbook of Experimental Phenomenology: Visual Perception of Shape, Space and Appearance.* Oxford: Wiley-Blackwell, 2013; 369–394.

[32] Helmholtz, H. von. *Helmholtz's Treatise on Physiological Optics.* Reprinted from the 1924–1925 edition, **Vol. 2**. Bristol: Thoemmes Press, 2000.

[33] Hurlbert, A. Colour constancy. *Current Biol.* 2007; 17(21): R906–R907.

[34] Pizlo, Z. Perception viewed as an inverse problem. *Vision Res.* 2001; 41(24): 3145–3161.

[35] Poggio, T., and Koch, C. Ill-posed problems in early vision: From computational theory to analogue networks. *Proc. Roy. Soc. B: Biol. Sci.* 1985; 226(1244): 303–323.

[36] Bloj, M. G., Kersten, D., and Hurlbert, A. C. Perception of three-dimensional shape influences colour perception through mutual illumination. *Nature* 1999; 402(6764): 877–879.

[37] Hurlbert, A. C. The chromatic Mach card. In Shapiro, A. G., and Todorovic, D., eds. *The Oxford Compendium of Visual Illusions.* Oxford: Oxford University Press, 2017; 382–387.

[38] Maloney, L. T., and Zhang, H. Decision-theoretic models of visual perception and action. *Vision Res.* 2010; 50(23): 2362–2374.

[39] Lee, D., and Plataniotis, K. N. A taxonomy of color constancy and invariance algorithm. In Celebi, M. E., and Smolka, B., eds. *Advances in Low-Level Color Image Processing.* Dordrecht: Springer, 2014; 55–94.

[40] Brainard, D. H., and Hurlbert, A. C. Colour vision: Understanding #TheDress. *Current Biol.* 2015; 25(13): R551–R554.

[41] BBC, *Colour: The Spectrum of Science*. 2015.

[42] Witzel, C., Racey, C., and O'Regan, J. K. The most reasonable explanation of 'the dress': Implicit assumptions about illumination. *J. Vision* 2017; 17(2).

[43] Lafer-Sousa, R., Hermann, K. L., and Conway, B. R. Striking individual differences in color perception uncovered by 'The Dress' photograph. *Current Biol.* 2015; 25(13): R545–R546.

[44] Gegenfurtner, K. R., Bloj, M., and Toscani, M. The many colours of 'the dress'. *Current Biol.* 2015; 25(13): R543–R544.

[45] Uchikawa, K., Morimoto, T., and Matsumoto, T. Understanding individual differences in color appearance of '#TheDress' based on the optimal color hypothesis. *J. Vision* 2017; 17(8).

[46] Wallisch, P. Illumination assumptions account for individual differences in the perceptual interpretation of a profoundly ambiguous stimulus in the color domain: 'The dress'. *J. Vision* 2017; 17(4).

[47] Aston, S., and Hurlbert, A. C. What #theDress reveals about the role of illumination priors in color perception and color constancy. *J. Vision* 2017; 17(9): 1–18.

[48] Toscani, M., Gegenfurtner, K. R., and Doerschner, K. Differences in illumination estimation in #thedress. *J. Vision* 2017; 17(1): 1–14.

[49] Hering, E. *Outlines of a Theory of the Light Sense*. Cambridge, MA: Harvard University Press, 1964 (first published 1874).

[50] Mollon, J. D. 'Cherries among the leaves': The evolutionary origins of color vision. In Davis, S., ed. *Color Perception: Philosophical, Psychological, Artistic and Computational Perspectives*. Oxford: Oxford University Press, 2000; 10–30.

[51] Jacobs, G. H. Primate photopigments and primate color vision. *Proc. National Acad. Sci. USA* 1996; 93(2): 577–581.

[52] Dominy, N. J., and Lucas, P. W. Ecological importance of trichromatic vision to primates. *Nature* 2001; 410(6826): 363–366.

[53] Lucas, P. W., Darvell, B. W., Lee, P. K., Yuen, T. D., and Choong, M. F. Colour cues for leaf food selection by long-tailed macaques (*Macaca fascicularis*) with a new suggestion for the evolution of trichromatic colour vision. *Folia Primatol.* 1998; 69(3): 139–152.

[54] Sumner, P., and Mollon, J. D. Catarrhine photopigments are optimized for detecting targets against a foliage background. *J. Exp. Biol.* 2000; 203(13): 1963–1986.

[55] Changizi, M. A., Zhang, Q., and Shimojo, S. Bare skin, blood and the evolution of primate colour vision. *Biol. Lett.* 2006; 2(**2**): 217–221.

[56] Fernandez, A. A., and Morris, M. R. Sexual selection and trichromatic color vision in primates: Statistical support for the preexisting-bias hypothesis. *Amer. Naturalist* 2007; 170(1): 10–20.

[57] Neitz, J., and Neitz, M. Evolution of the circuitry for conscious color vision in primates. *Eye* 2016; 31(2): 286–300.

[58] Walmsley, L., Hanna, L., Mouland, J., Martial, F., West, A. et al. Colour as a Signal for Entraining the Mammalian Circadian Clock. *PLoS Biol.* 2015; 13(4).

[59] Lafer-Sousa, R., Liu, Y. O., Lafer-Sousa, L., Wiest, M. C., and Conway, B. R. Color tuning in alert macaque V1 assessed with fMRI and single-unit recording shows a bias toward daylight colors. *J. Opt. Soc. Amer. A: Optics Image Sci. Vision* 2012; 29(5): 657–670.

[60] Nathans, J. The evolution and physiology of human color vision: Insights from molecular genetic studies of visual pigments. *Neuron* 1999; 24(2): 299–312.

[61] Buchsbaum, G., and Gottschalk, A. Trichromacy, opponent colours coding and optimum color information transmission in the retina. *Proc. Roy. Soc. B: Biol. Sci.* 1983; 220(1218): 89–113.

[62] Milojevic, Z., Ennis, R., Toscani, M., and Gegenfurtner, K. R. Categorizing natural color distributions. *Vision Res.* 2018; 151: 18–30.

[63] Ling, Y., Vurro, M., and Hurlbert, A. C. Surface chromaticity distributions of natural objects under changing illumination. In *Proceedings of the 4th European Conference on Colour in Graphics, Imaging and Vision.* Terrassa: Society for Imaging Science and Technology, 2008; 263–267.

[64] Gibson, E., Futrell, R., Jara-Ettinger, J., Mahowald, K., Bergen, L. et al. Color naming across languages reflects color use. *Proc. National Acad. Sci. USA* 2017; 114(40): 10785–10790.

[65] Webster, M. A., Miyahara, E., Malkoc, G., and Raker, V. E. Variations in normal color vision. II. Unique hues. *J. Opt. Soc. Amer. A: Optics Image Sci. Vision* 2000; 17(9): 1545–1555.

[66] Conway, B. R. Color signals through dorsal and ventral visual pathways. *Visual Neurosci.* 2014; 31(2): 197–209.

[67] Chevreul, M. E. *The Laws of Contrast of Colour: And Their Application to the Arts.* London: George Rutledge and Sons, 1868.

[68] van Gogh, V., *Letter to Theo (533), Arles, 8 September 1888.* In Powell, E., ed. *The Letters of Vincent van Gogh to His Brother and Others 1872–1890.* London: Constable & Robinson Ltd, 2003; 533.

[69] Veen, L. Piet Mondrian on the principles of neo-plasticism. *Int. J. Art Art History* 2017; 5(2): 1–12.

[70] Hurlbert, A., and Wolf, K. Color contrast: A contributory mechanism to color constancy. In Heywood, C. A., Milner, A. D., and Blakemore, C., eds. *Roots of Visual Awareness: A Festschrift in Honor of Alan Cowey.* Amsterdam: Elsevier, 2003, 147–160.

[71] Brown, R. O., and MacLeod, D. I. A. Color appearance depends on the variance of surround colors. *Current Biol.* 1997; 7(11): 844–849.

[72] Hurlbert, A. C. Computational models of colour constancy. In Walsh, V., and Kulikowski, J., eds. *Perceptual Constancy: Why Things Look as They Do.* Cambridge: Cambridge University Press, 1998; 283–321.

[73] Lee, B. B., Dacey, D. M., Smith, V. C., and Pokorny, J. Horizontal cells reveal cone type-specific adaptation in primate retina. *Proc. National Acad. Sci. USA* 1999; 96(25): 14611–14616.

[74] Conway, B. R. Spatial structure of cone inputs to color cells in alert macaque primary visual cortex (V-1). *J. Neurosci.* 2001; 21(8): 2768–2783.

[75] Smithson, H. E. Sensory, computational and cognitive components of human colour constancy. *Philos. Trans. Roy. Soc. B: Biol. Sci.* 2005; 360(1458): 1329–1346.

[76] Foster, D. H., Craven, B. J., and Sale, E. R. H. Immediate color constancy. *Ophthalm. Physiol. Optics* 1992; 12(2): 157–160.

[77] Barbur, J. L., and Spang, K. Colour constancy and conscious perception of changes of illuminant. *Neuropsychologia* 2008; 46(3): 853–863.

[78] Norman, L. J., Akins, K., Heywood, C. A., and Kentridge, R. W. Color constancy for an unseen surface. *Current Biol.* 2014; 24(23): 2822–2826.

[79] Crichton, S. O. J., Pichat, J., Mackiewicz, M., Tian, G.-Y., and Hurlbert, A. Skin chromaticity gamuts for illumination recovery. In *6th European Conference on Colour in Graphics, Imaging and Vision.* Amsterdam: The Society for Imaging Science and Technology, 2012; 266–271.

[80] Hansen, T., Olkkonen, M., Walter, S., and Gegenfurtner, K. R. Memory modulates color appearance. *Nature Neurosci.* 2006; 9(11): 1367–1368.

[81] Hurlbert, A., van Zuijlen M., Spoiala, C., and Wijntjes, M. Colour determines perceived circadian phase in visual art. In Abstracts from the 6th Visual Science of Art Conference (VSAC), Trieste, Italy, August 24th–26th, 2018. *Art & Perception* 2018; 6(4): 198.

[82] Land, E. H., and McCann, J. J. Lightness and retinex theory. *J. Opt. Soc. Amer.* 1971; 61(1): 1–11.

[83] Arend, L., and Reeves, A. Simultaneous color constancy. *J. Opt. Soc. Amer. A: Optics Image Sci. Vision* 1986; 3(10): 1743–1751.

[84] Foster, D. H. Color constancy. *Vision Res.* 2011; 51(7): 674–700.

[85] Rewald, J. *The History of Impressionism.* London: Secker and Warburg, 1973.

[86] Bosten, J. M., Beer, R. D., and MacLeod, D. I. A. What is white? *J. Vision* 2015; 15(16).

[87] Chauhan, T., Perales, E., Xiao, K., Hird, E., Karatzas, D., and Wuerger, S. The achromatic locus: Effect of navigation direction in color space. *J. Vision* 2014; 14(1:25): 1–11.

[88] Winkler, A. D., Spillmann, L., Werner, J. S., and Webster, M. A. Asymmetries in blue–yellow color perception and the color of 'the dress'. *Current Biol.* 2015; 25(13): R547–R548.

[89] Weiss, D., Witzel, C., and Gegenfurtner, K. Determinants of colour constancy and the blue bias. *I-Perception* 2017; 8(6).

[90] Kuriki, I. The loci of achromatic points in a real environment under various illuminant chromaticities. *Vision Res.* 2006; 46(19): 3055–3066.

[91] Gupta, G., Gross, N., Pastilha, R., and Hurlbert, A. The time course of color constancy by achromatic adjustment in immersive illumination: What looks white under coloured lights? *bioRxiv* 2020.

[92] Pearce, B., Crichton, S., Mackiewicz, M., Finlayson, G. D., and Hurlbert, A. Chromatic illumination discrimination ability reveals that human colour constancy is optimised for blue daylight illuminations. *PLoS ONE* 2014; 9(2): e87989.

[93] Radonjić, A., Pearce, B., Aston, S., Krieger, A., Dubin, H. et al. Illumination discrimination in real and simulated scenes. *J. Vision* 2016; 16(11): 1–18.

[94] Aston, S., Radonjić, A., Brainard, D. H., and Hurlbert, A. C. Illumination discrimination for chromatically biased illuminations: Implications for color constancy. *J. Vision* 2019; 19(3): 1–23.

[95] Hamilton, J., *Turner: A Life.* London: Hodder and Stoughton, 1997.

4 Science, Vision, Perspective

CARLO ROVELLI

Introduction

Many scientific ideas, including in modern physics, have a key visual component. Images are not trivial metaphors for popular science: they are a powerful scientific tool, which propels and nourishes the best science. But the relation between vision and science runs deeper: there is a telling similarity between how science works and how the brain performs vision. Recognising it sheds light on the nature of scientific knowledge. Finally, the proximity between vision and science illuminates also a notion playing an increasingly general role in modern physics: perspective.

Visual Imagery in Science

Major scientific discoveries have been suggested by, inspired by, or can be condensed into, a simple picture.

Here are a few famous examples. In Copernicus's great book, the *De Revolutionibus*, the drawing of the Solar System with the Sun at the centre (Figure 4.1) encapsulates the content of the heavily technical book into a single image: the Sun at the centre, and the Earth, with its moon, that rotates around it.

Isaac Newton's great discovery, that falling on Earth and orbiting are two instances of the same phenomenon, is suggested, in fact virtually demonstrated, by a single picture (Figure 4.2), which appears in Newton's *System of the World*. The picture depicts the Earth and a high mountain from which a stone is ideally thrown with increasing force. The picture shows eloquently that, when the force is sufficient, the stone comes back to the thrower: it is in orbit.

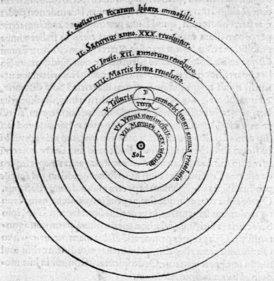

NICOLAI COPERNICI

net, in quo terram cum orbe lunari tanquam epicyclo contineri diximus. Quinto loco Venus nono mense reducitur. Sextum deniq; locum Mercurius tenet, octuaginta dierum spacio circu currens. In medio uero omnium residet Sol. Quis enim in hoc

pulcherimo templo lampadem hanc in alio uel meliori loco po neret, quàm unde totum simul possit illuminare? Siquidem non inepte quidam lucernam mundi, alij mentem, alij rectorem uo= cant. Trimegistus uisibilem Deum, Sophoclis Electra intuentē omnia. Ita profecto tanquam in solio re gali Sol residens circum agentem gubernat Astrorum familiam. Tellus quocq; minime fraudatur lunari ministerio, sed ut Aristoteles de animalibus ait, maximā Luna cū terra cognationē habet. Concipit interea à Sole terra, & impregnatur annuo partu. Inuenimus igitur sub hac

FIGURE 4.1 The page of Copernicus's book with the image of the Solar System seen from the exterior. PHOTOS.com/Getty Images Plus via Getty Images.

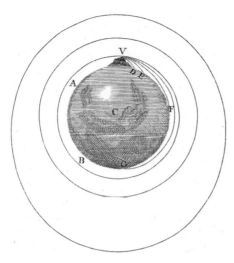

FIGURE 4.2 A stone thrown horizontally with increasing strength
ends up respectively at the points D, E, F, B, A and then 'falls' along a circular orbit.
This drawing by Isaac Newton shows that falling and orbiting are the same
phenomenon. From Sir Isaac Newton (1642–1727), image from *Philosophiae Naturalis
Principia Mathematica*. Reproduced by kind permission of the Syndics of Cambridge
University Library.

The famous 'tree of life' scheme in Charles Darwin's notebook captures
his initial intuition about the family relation between all species on Earth
in a simple sketch (Figure 4.3).

The helicoidal structure of DNA (discovered with Rosalind Franklin)
was immediately realised visually by means of a three-dimensional model
(Figure 4.4) by Francis Crick and James Watson, in order to illustrate it
to everybody.

Modern fundamental physics is based on the notion of *field*. The
concept of field is the basic theoretical tool both in gravitational physics
and in particle physics. A field is an entity diffused in space. The first to
conceive the idea of a field was Michael Faraday. In his book *Experimental
Researches in Electricity* he candidly gives us a picture (Figure 4.5) indicat-
ing the source of his intuition of what a (magnetic) field is: an invisible
bundle of lines filling space around a magnetic bar.

These, like many others, are cases where a truly major scientific
advance can be synthesised into a single image and was probably sparkled
by this image.

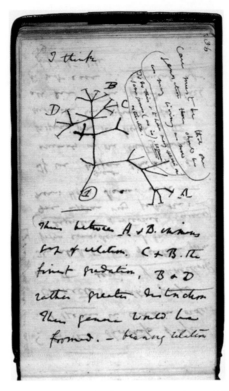

FIGURE 4.3 The drawing of the 'tree of life' in Charles Darwin's notebook entry in 1836: his initial insight into his major discovery. Reproduced by kind permission of the Syndics of Cambridge University Library. Pg.6.tif (Rel.f.1b.48, p.6).

I do not think that these representations of great discoveries by a single image are a way to trivialise them. Discoveries require cumulated knowledge, accurate and painstaking observation, mathematics, rational thinking, trial and error, and all that. But the creative moment that sparks a discovery is often a single visual image. Once proven effective, the core of the discovery can still be encapsulated into that single image.

Einstein's visual imagination was legendary. He has written that this visual imagination was instrumental in leading his intuition and his momentous discoveries. He wondered what he would see riding a light wave, visualised the granular structure of light before quantum theory,

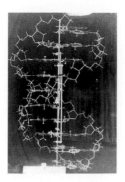

FIGURE 4.4 The six-foot-tall metal DNA model made by James Watson and Francis Crick in 1953. Courtesy of Cold Spring Harbor Laboratory.

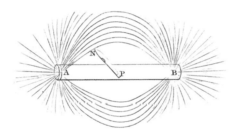

FIGURE 4.5 Michael Faraday's picture of magnetic field lines: the intuition from which stems the modern notion of field.

imagined the curvature of spacetime before learning the maths to describe it in equations, and so on [1].

Already in antiquity visual aspects dominated science. Anaximander 'saw' that the Earth floated in the middle of the sky [2], Aristarchus and Hipparchus used geometry, hence pictures, to correctly estimate the size and distance of celestial bodies (Figure 4.6).

In my own field of research, quantum gravity, a powerful visual image that has driven and is still driving research is the granular nature of space, due to the quantum nature of gravity. Two individual grains (or quanta) of space can be adjacent to one another. This adjacency is the basis of spatial relations. The quanta of space and their adjacency relation form a network, a graph, which is called a 'spin network' (Figure 4.7) in the jargon of the theory.

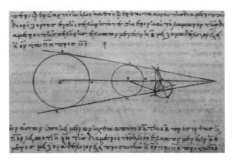

FIGURE 4.6 Greek parchment of the tenth century, reporting early astronomical calculation on the Sun–Earth–Moon system, using a visual representation of a lunar eclipse from a side perspective.

FIGURE 4.7 A spin network. The nodes of the graph represent the individual quanta of space. They are not immersed in a space: they themselves weave space.

And so on.

Not every scientist has such a strongly visual approach. A well-known counterexample is Paul Dirac, who, together with Einstein, was one of the greatest physicists of the twentieth century. Dirac was famous for the very abstract character of his science. If Dirac visualised his theoretical reasoning, he did so in extremely abstract form. The elegant equation that carries his name, for instance, which is used today to predict the motion of electrons and quarks,

$$i\hbar\gamma^\mu\partial_\mu\psi = mc\psi,$$

notoriously defies simple visualisation.

If I am allowed a personal note, I have had the good fortune of leisurely observing two great scientists working together: Roger Penrose and Ted

Newman. Ted's office was next to mine in Pittsburgh, and I used to sneak in and listen when Roger visited. The conversation between the two was fascinating for many reasons. One of these was that the two of them seemed to speak in different tongues: Penrose is among the most visual and geometrical among the modern scientists, while Newman is a bit like Dirac: he thinks in equations, not pictures. And still, although painfully after much confusion, they would end up understanding one another. They have written together beautiful and important pages of the mathematical physics of general relativity. I can represent what I was watching (in the spirit of this chapter) with an image (Figure 4.8).

Modern physics is often said to be hard to visualise. But this is simply lack of familiarity. From Einstein's spacetime waves to the atomic quantum orbitals, from the expanding Universe to the black holes bouncing into white holes, modern physics is in reality a feast of visualisations.

FIGURE 4.8 Roger Penrose (left) and Ted Newman (right), each with a representation of their different ways of describing the same phenomena: visual and geometrical (Penrose) versus algebraic (Newman).

The relationship between science and vision, however, runs deeper than this simple observation. Science and vision may work in a remarkably similar manner. This is discussed in the next section.

Science and Vision

In science, we have theories. Theories allow us to predict aspects of phenomena. We keep collecting new inputs, with observations and measurements. The theories allow us to predict data and observations. While these predictions are sufficiently accurate, the theory is considered good and allows us to better navigate the world. When the predictions are contradicted by new data, we have new relevant information, which motivates us to modify the theory. A successful modification of the theory makes us able to take the new data into account and increases the predictive power. Science is the historical, dynamical, process of increasingly refining theories whenever new data contradict them.

The role of data contradicting predictions of current theories is so important in science that the philosopher Karl Popper indicated it as the demarcation criterion for science [3]. That is, he suggested that one should call a theory 'scientific' *only* if it can in principle be contradicted ('falsified', in his jargon) by new data. The idea is that the effectiveness of scientific knowledge relies precisely on its constant openness to being put in question, because this process allows constant improvement of knowledge, which is the mark of science.

The interpretation of science as an *historical* process of growth of knowledge, alternating successful predictive theories and novel discrepancies (or accumulation of discrepancies) leading to modifications of the theoretical corpus, has been developed by Thomas Kuhn [4] (perhaps over-emphasising discontinuity). The role of Bayesian probabilistic thinking in science and in particular the updating of priors under Bayesian 'dis-confirmation', namely gradually adjusting a hypothesis to better fit the data, was clarified by Bruno de Finetti [5] and developed by Frank Ramsey [6], and is increasingly recognised in the recent philosophy of science literature as a core characterisation of the way science actually works.

Animals' vision, on the other hand, is a complex natural phenomenon happening in the eyes and the brain. Light passing though the pupil interacts with a fluctuating sea of neural signals and ends up providing the brain with a codification and an understanding of the outside world, which is useful to navigate this world in such a way as to maximise the chance of survival.

How this concretely happens in the brain is currently under intense investigation by the neurosciences. Among the most interesting results of these investigations is increasing evidence that the relevant flow of information is not just 'upward' from the eyes to the brain's upper layers, as a naive model of vision might suggest. Rather, the brain uses long- and short-time memory to constantly *predict* sensory inputs; the 'upward' flow codes only the discrepancies between these predictions and new inputs.

In more detail, the brain has a layered structure; each layer receives probabilistic predictions from the upper layer and inputs from the lower one, compares these inputs, and signals discrepancies upwards. The system constantly tunes itself to minimise discrepancies, ameliorating prediction. The upper layers deal with increasingly general and abstract aspects of the world. Trivialising: if lower levels can predict where a tossed coin will land, an upper level might predict which of the passing characters will kindly toss a coin to the street singer. Lower levels encode a model of the trajectory of the coin; upper levels a complex assessment of relevant aspects of the entire visual scene. This is called the predictive processing, or predictive coding, framework. For a review, see for example [7]. It originated from a back and forward between studies in artificial intelligence and brain structure.

According to this model, the world we 'see' is what the brain predicts, rather than directly the input from the retina. In a sense, the brain imagines the world, and constantly corrects the image against incoming inputs. In the words of the nineteenth-century French philosopher Hippolyte Taine, 'external perception is an internal dream which proves to be in harmony with external things; and instead of calling hallucination a false external perception, we must call external perception true hallucination' [8]. The Darwin lecture by the British psychiatrist Paul Fletcher (from whom I have taken the above quote), Chapter 2 in this volume, brilliantly reviews these advances in neuroscience and their relevance to our understanding of hallucination.

The similarity between this account of vision as predictive processing and the understanding of how science works recalled above is remarkable. There is first of all an emphasis on prediction in both cases. But prediction has a double role: it increases our ability to navigate the world, ultimately motivating the entire process, and it constantly tests the theory. Then, both in vision and in science, the key is the error-correction process, which is based on focusing on any *discrepancy* between predictions and new inputs. Overall, this is a predictive world-picture, constantly checked and updated against incoming input.

The difference between vision and science is mostly a difference in time scale and in the elaborating tool. In vision, the updating process can happen in a fraction of a second and takes place within a single brain. In science, the updating process happens at a much slower pace, on a time scale of perhaps years, decades, or centuries, and occurs in the collective cultural exchange space of a scientific community.

To be sure, this similarity could be a mirage: perhaps we are misinterpreting the brain, projecting our own cultural methodology onto it. But, if the analogy is true, it is instructive for a number of reasons.

First, an aspect of the process of vision in the brain is its hierarchical organisation in layers. The process of downward prediction, detection of discrepancies with the input, and upward updating of priors, happens at each layer in the brain, generating, as we move upwards, increasingly general codification of the sensory inputs' structure. The upper levels capture increasingly deep and general regularities in the inputs.

Science, then, can be viewed – in a sense – as nothing more than one further upper level of the same process: a layer that works slower, does not happen in a single brain, involves the full structure of the cultural exchange of information, and codifies very general regularities of nature in the form of scientific laws and theories. But what it does is nothing else than Bayesian updating of priors, on the basis of detected discrepancies, as each layer of the brain does.

Second, this understanding of the nature of science reinforces its fully naturalistic interpretation, in the sense of Quine [9]. Science is a refined version of what we normally do in life, enhanced by the remarkable collaborative attitude of our species.

Indeed, the contiguity and structural similarity between vision and science suggests that we can consider science as a sort of extension of vision; hence, so to say, a natural expansion of common-sense knowledge, and not as something alternative to it. Contrary to what is often stated, there is then no hard separation between the 'scientific view' and the 'common view' of reality. They are more or less effective levels of modelling the world.

Of course, scientific discoveries sometimes challenge common sense. We feel the Earth to be at the centre of a rotating sky; while we learn from science that it actually spins under a still sky. But challenging current knowledge is not specific to science: it is ubiquitous in any learning process. The child who at a certain age learns that their small village is not the centre of the world is going through the same kind of learning process humanity as a whole underwent in learning, with Copernicus, that the Earth is not the centre of the Universe. Both are learning a larger and more effective picture, which extends the previous one and partially contradicts it. Science is intelligent continuation of common-sense learning.

Science can thus be seen as a fully natural enterprise. It is not a purely rational construct aiming to freeing us from all prejudices and giving us (approximations to) certitude, but rather a continuation and a refinement of our natural adaptation tools, which exploits the collaborative instinct of primates like us: an extension of the natural activity of our brain on the basis of prediction and error correction.

Returning now to the parallel between vision and the evolution of a scientific hypothesis, how can vision use the light entering the pupil to predict what comes next? The answer is that there are patterns in the incoming light. In between two small nearby blue patches of sky, the sky is very probably blue. A yellow stone stably releases yellow photons, all coming towards the eye from the same direction. A pattern is any feature in the data – a structure – that allows data compression, and hence permits prediction, both in space and in time.

Sensory experience is richly structured in patterns. (Why? Depending on one's philosophy, this can be taken as an argument for realism, or for

God's goodness, or just as a general aspect of the world we experience; or possibly something else. But the fact itself remains, and it is the fact that permits our knowledge of the world.) When these patterns are recognised by the brain, they allow data to be *organised*, and organised data allow partial predictions. The neural algorithm underlying vision is thus pattern recognition, namely organisation of the data-inputs into increasingly general patterns by increasingly higher layers of the brain.

Science, similarly, is a (collective) search for patterns of regularity in the observed world. These are expressed in the form of scientific laws and updated when contradicted by new data. The individual brain and the collective enterprise called culture are both constantly searching for increasingly effective patterns organising experience, continuing an activity that increased evolutionary chances.

The idea that this *organisational* effort is the key notion for understanding knowledge acquisition, as well as our activity itself, goes back to Alexander Bogdanov [10], whose important influence on current thinking can be traced to his legacy in artificial intelligence – which has inspired the predictive processing framework – via his direct influence on early cybernetics [11] and systems theory [12]. Notice, in passing, that from this point of view entities are structures. But they are structures that are not necessarily stable. First, they can be structures in time. More importantly, they evolve as we learn.

To conclude this first section: vision and science may work in the same manner, namely by using discrepancies to constantly revise and update a predictive tool that scouts for patterns for organising data. Science can be viewed as an upper level of vision, a fully natural extension of normal brain activity.

In the next section, I explore a further and deeper aspect of the similarity between vision and science: the role of perspective.

Perspective

Consider again the pictures presented in the first section of this chapter. Each of them, just like innumerable others in science, can be described as the image of a phenomenon from a new perspective: an efficacious change of perspective. A scientific discovery is often the discovery of a better 'view', a better 'perspective', on a certain phenomenon or family of phenomena.

Anaximander, Aristarchus, and Newton 'saw' the Earth from afar, Copernicus saw the Solar System from outside it. Darwin 'saw' the path traced by the evolving species over hundreds of millennia. Franklin, Crick, and Watson 'saw' the microstructure of a molecule as if they could zoom in. Quantum gravity is the attempt to zoom in further, all the way to the microstructure of space. And so on.

In vision, the brain recognises an object (or perhaps *constructs* it, depending on one's philosophy), for instance an elephant, from a bundle of signals. A moment of reflection shows how incredibly fractured and prima facie incoherent is the actual visual input from the retina, from which an object and its location are recognised. For an illuminating discussion on the complexity of vision in this regard see Ismael's 'Do you see space?' [13].[1] A crucial aspect of this process is the ability to recognise the object even if it moves or rotates, if external conditions (say ambient light) change, and especially if the spatial position of the body of the viewer, hence the location of her or his eyes, changes. This is the issue of *perspective*. The same object, the same process, the same scene, can be viewed from different perspectives: different angle, different distance, different resolution, and so on.

Perspective Has Two Separate Sides, a Bit in Tension with One Another

On the one hand, the brain aims at recognising something as *the same*, when the viewer changes perspective on it. Consequently, the brain constantly computes how the appearance of objects would change, if the object or the viewer's body moved. The image of the world that the brain holds allows it to combine past knowledge and current inputs, to predict what it would see, if it saw the same object from a different point of view.

On the other hand, a change of perspective can provide *new* relevant information on the same object. No object is well coded in our brain until it has been seen from a variety of perspectives. If we see an elephant only from the back, we may never know it has a long nose. If we see a forest

---

[1] I thank Jenann Ismael for letting me read a preliminary draft.

only from the other side of the valley, we may never know the moss on the trunks of the trees, or the colour of the ladybirds flying around in it.

Change of perspective is thus a crucial aspect of our manner of apprehension of the very complex three-dimensional world in which we happen to live, and the vast variety of what is in it. We learn about the world by actively *moving in it* and repeatedly *changing the perspective* from which we see things.

Both these aspects of perspective play an important role in science, and considering them sheds light on the way science works.

Let me first consider the ability of predicting what an object would look like, from a point of view we have not yet taken.

Our concrete possibilities of changing perspective are limited. For a long time, for instance, we could not fly outside the Earth to take a look at its shape from afar. We could not zoom into molecular scales and directly see the shapes of molecules. We could not watch a black hole from close enough to figure out what it actually looks like.

But the brain is designed precisely to maximally exploit any available hint to effectively compute and predict how things would look from a new perspective. So much so, that Anaximander, Aristarchus, and Copernicus largely predicted what astronauts would see looking at the Earth from the Moon: a rotating ball (Figure 4.9). Chemists and physicists correctly predicted the shapes of molecules, which were only much later observed via atomic microscopes (Figure 4.10). Relativists correctly predicted how a black hole would appear if we could see it. We finally saw one, this year, in the first image obtained with the Event Horizon Telescope (Figure 4.11) [14].

A beautiful geometrical demonstration by Hipparchus in the second century before our era, to compute the distance of the Moon, starts by asking what one would see if one could fly to the tip of the cone formed by the shadow of the Earth. (One would see the Earth perfectly hiding the Sun, which is very far away, showing that the angle at this tip is the same as the known angle under which we see the Sun.) This is a beautiful example of science imagining and using a radical change of perspective.

And so on. These and innumerable others are examples of the ability of science to predict the result of change of perspective, as the brain does routinely in vision, but in a far more unfamiliar context.

FIGURE 4.9 The Earth as seen from the perspective of the Apollo 11 astronauts. Courtesy of NASA.

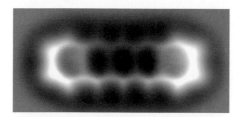

FIGURE 4.10 A molecule of pentacene, seen with an atomic-force-microscope image, from IBM Zurich.

FIGURE 4.11 Photo of the large black hole at the centre of the M87 galaxy, taken with the Event Horizon telescope. Credit: EHT Collaboration.

The more the process of widening perspectives and the prediction ability extends, the harder and slower it becomes: instead of a rapid process in the brain, it becomes a slow process in the exchanges between thinkers, who use observations, data, imagination, arguments, and mathematics, to make the predictions and anticipation process more powerful. Again, vision evolves into science as a collective enterprise.

But it is the second aspect of perspective that is more telling, with regard to science: we *learn* about something by shifting our point of view on it, by realising how it looks from a different perspective. A large number of scientific advances can indeed be simply interpreted as the discovery of a new and more encompassing *perspective* on a phenomenon. Which is to say, recognising that the previously held perspective was a limited one.

Resolving scientific puzzles by recognising a perspectival aspect of our data is indeed an ancient and noble tradition in science. A prototypical example is (Aristarchus' and) Copernicus's answer to the question 'Why does the entire Cosmos – Sun, Moon and stars – turn once a day around us?' The answer, of course, is that it doesn't: the apparent daily motion of the sky is a *perspectival* effect due the fact that we happen to dwell on a spinning stone, the Earth. We recognise that a major phenomenon observed in the inputs the eye receives depends on specific aspects of the point of view of the eye.

Notice that perspectival explanations like this one are subtle. They involve the point of view of the 'observer', but there is nothing subjectivist, mentalist, or idealistic in them. The fact is simply that the sky turns 'with respect to' a spinning rock; and this is a fact that has no subjectivist, mentalist, idealistic implication whatsoever. It is perspectival in the sense that it pertains to *two* physical entities (the sky and the planet) and their *relation*, not just to the observed phenomenon alone (the sky).

Physicists tend to slide into a misleading subjectivist language: the object with respect to which relative velocity is taken is routinely called the 'observer', which suggests that it needs to have a mind. But in physics 'relative to' does not mean relative to a subject. It only means that two objects, rather than a single one, are involved. The distance of a street sign from a road crossing, for instance, is a quantity 'relative to the road crossing': this does not mean we are assuming subjectivity in road crossings. Perspectivalism is fully and strictly within the boundaries of physicalism.

In modern physics, such a perspectival reasoning is playing an increasingly extensive role. To illustrate this claim, consider the following examples.

Velocity Is Perspectival

The Copernican revolution raised an obvious problem: we do not feel the fast motion of our planet that its new perspective seems to reveal. The solution, foreseen by Galileo, is one of the pillars of classical physics. We must take into account that velocity itself is a *relational*, or *perspectival*, concept. That is, any assessment of velocity of an object is always (explicitly or implicitly) *relative to another object*. In the physicists' jargon, it is 'relative to a reference frame'. A 'reference frame' is an ideal collection of objects that do not move with respect to one another. This brilliant solution amounts to the discovery of a new level of perspective: the sensory inputs we receive do not depend only on the position of the viewer, but also on her or his *motion*. Velocity is a quantity that characterises the relation between *two* physical bodies: it is always a velocity of one object from the perspective of another.

The key point is not so much that the observed velocity depends on the observer's motion, which is pretty trivial. (If a mother says 'stay still' to her child on a train, she does not mean that the child should jump out of the train and remain still with respect to the ground.) The key point is that there is no *preferential* state of motion. There is no preferred reference frame from which to see the world. All perspectives are physically equivalent.

There is no *preferred* perspective. This implies that the best generalisation for understanding the world must be sought by understanding the *relations* between perspectives, rather than trying to explain perspective away, translating inputs to a preferred perspective.

Elementary Particles Are Perspectival

If we disregard gravity altogether, modern fundamental physical theory (quantum field theory) predicts that the elementary constituents of nature, which are mathematically represented as quantum fields, manifest

Carlo Rovelli

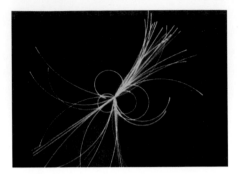

FIGURE 4.12 A photo of the tracks of elementary particles emerging from a collision at CERN: matter seems to be constituted by elementary particles. But it is not: some of these particles may not exist for a freely falling detector. Pier Marco Tacca/ Getty Images News via Getty Images.

themselves in the form of *particles*. This matches very well with the observations in experimental high-energy physics, or 'particle' physics. For instance, the detectors at CERN, in Switzerland, take pictures of tracks of elementary particles such as electrons or muons (Figure 4.12).

In turn, this fact can be taken as support for a particle ontology: matter is made by elementary particles. In a finite region of space, in a small time interval, we can say 'how many particles' are there. Such measurements are routine in the particle detectors at CERN.

However, if we take (classical) gravity into account, something dramatic happen. The theoretical framework that takes classical gravity into account is called 'quantum field theory on curved spacetimes'. It is the most general theoretical framework currently considered plausible by the theoretical physics community as a whole. (It still neglects *quantum* gravity, on which there is no consensus yet.) Taking classical gravity into account, something remarkable follows: the number of particles detected in the *same* region of space at the *same* time by the *same* kind of detector depends on the *acceleration* of the detector. For instance in an 'empty' region of space where a fixed detector would detect no particles, a detector in accelerated motion will 'see' particles. This is called the Unruh effect, from the Canadian physicist Bill Unruh, who understood it first [15].

To fully appreciate the relevance of this (theoretical) discovery, observe that a 'non-accelerating' detector in general relativistic parlance is a detector that is freely falling. The CERN 'still' detectors are therefore accelerating (in this sense): hence some of the particles they see do not exist, insofar as a freely falling detector is concerned.

This is a subtle effect, but its implications are heavy: they demolish the possibility of taking a realist particle ontology seriously. The number of particles is a *perspectival* quantity: it depends on the state of motion of whatever is counting them. It is hard to maintain a realist particle ontology, if the number of particles in a region is not a property of the matter in the region, but depends on the system counting these particles. Again, perspective comes to the front in contemporary theoretical physics.

Gauge Theories

A third important example of perspectivalism in modern physics is provided by gauge theories. Gauge theories are the particular form of field theories that describe particle physics. They are characterised by a redundancy in their mathematics: different mathematical solutions of the equations of the theory are interpreted as representing the same physical state of affairs. Trying to get rid of this (apparently insane) mathematical redundancy ends up complicating the mathematics beyond control. Why is this so?

It is because gauge theories are not about the physics at individual points in space: they are about relations between the physics in different points [16].

This can be illustrated with an elementary example, taken from the simplest gauge theory: electromagnetism. To describe electrical phenomena we use the notion of electric potential, measured in volts. The difference between the electric potential in the two holes of an electric plug is 220 V. The difference between the electric potential in the two poles of a battery is 12 V. And so on. Why do we always talk about 'differences' in electric potential, rather than just giving the value of the potential of each single hole of the plug, or each single pole of the battery?

Because the electric potential of a single point is a physically meaningless concept. What matters about potential is only differences between *two* points. It is this and only this that has physical consequences. Electric potential is not a property of an object alone: it is a property of an object with respect to another object.

We can (and often do) choose a single object (say the negative pole of the battery, or the Earth) and assign the value *zero* to its electric potential. Doing so amounts to evaluating the potential of anything else *with respect to this reference object*, or, in an extended sense, *from the perspective* of the reference object. If we then add a constant value to all potentials, we are describing the same physics. We are changing perspective: viewing potentials from a different reference object point of view.

Quantum Theory

There is no agreement among scientists (or philosophers) on what to make of the strangeness of quantum mechanics. But two of the most discussed interpretations of quantum mechanics make sense of the theory by adding 'a new level of indexicality' to our description of the world. These are the many-worlds [17] interpretation and the relational [18] interpretation.

The first interprets the quantum state realistically and reduces quantum indeterminism to *perspective* with respect to the 'branch' of the wave function where the observer happens to be.

The second does not interpret the quantum state realistically. It interprets realistically the values of variables instantiated in interactions. But it assumes that these values are instantiated only *with respect to* the interacting systems themselves, and not with respect to a third system.

The first adds a 'branch' indexicality, the second a system indexicality. Both resolve the puzzles of quantum mechanics by recognising a *perspectival* aspect of physical variables.

The Arrow of Time

Finally, one can speculate that even the arrow of time, namely the manifest difference between the past and the future, might be a perspectival phenomenon [19]. The reason is the following.

We know that we can trace any difference between past and future to the simple fact that in the remote past the Universe was in a 'special' (low-entropy) state. Entropy has since grown, driving all irreversible phenomena, namely giving time a preferred arrow. The notion that a state can be 'special', however, pertains to the macro-physics, not to the micro-physics. This is because, at the microscopic level, any state is equally special; while macroscopic states can be 'special' by virtue of their being realised by a small number of micro-states (namely having low-entropy). In turn, the macroscopic description of phenomena is determined by a relatively small number of 'macroscopic' variables, which are the (few) variables of the system that interact with a second 'observer' physical system. Hence any macroscopic description is perspectival: it is relative to a second system that happens to interact with the first via a given small set of macroscopic variables.

This opens the possibility that the initial low entropy of the Universe was not a property of the Universe by itself, but rather a property of the Universe with respect to a special particular class of subsystems, with which it interacts via a specific set of variables, which define a macroscopic state that was 'special' in the past. Perhaps we may understand the arrow of time in the same way as we understand the rotation of the sky: not as a property of the Universe at large, but as a property of the limited and peculiar perspective on it given to the small physical subsystems we belong to, and the variables of the rest of the Universe they interact with. This is a speculative idea; one I find especially intriguing.

All these examples show that modern physics has revealed a deep and multi-faceted perspectival structure of reality, which includes relativity, particle ontology, quantum mechanics, gauge theories, and perhaps even the arrow of time.

This ubiquity of perspectivalism has philosophical implications. The relational interpretation of quantum mechanics, in particular, makes sense of the strangeness of quantum theory at the price of assuming that the value of *any* variable that describes a physical system is *always* defined only relative to another physical system. That is, all variables are relational like velocity is. This may be hardly compatible with traditional substance metaphysics.[2]

On the other hand, the analogy with vision is illuminating: there is less and less 'vision from nowhere' in modern physics. Any vision implies a point of view, hence a perspective.

---

[2] Perhaps it requires a metaphysics of mutual dependence as in Nagarjuna's Mūlamadhyamaka-kārikā?

The best understanding of reality we can aim for will be achieved not by trying to transform perspective away, but by striving to understand the relations between perspectives.

Perspective is a key structure underlying vision: we do not make sense of the three-dimensional world we inhabit unless we consciously or unconsciously consider perspective, namely we take into account the (shifting) position from which we look at it. Science, I think, is undergoing the same learning process: we shall not make sense of the world we inhabit unless we fully grasp its relational and perspectival structure and the fact that we always access it from a particular point of view. I do not think we have fully done so yet.

Conclusion

Scientific intuition and scientific discovery are often driven by the visual. Science is not moving away from simple seeing. In fact, it can be seen as a natural extension of vision, since its logic is the same as that which one's brain probably uses in vision: detecting patterns stable enough to allow prediction and constantly testing and updating them. Like vision, science is a search for an increasingly effective world *view*, in a pretty literal sense.

Science sometimes teaches us that things are different from the way they look: reality *is not what it looks like*. The Earth is not flat, the Sun is not rotating around us, solid bodies are not compact, time is not universal . . . But it also teaches is that, all this notwithstanding, reality *still looks like something*. The Earth 'looks like a ball', the Solar system 'looks like a merry-go-around', a solid body at small scale is 'like the sky dotted with stars', and – as Einstein put it – spacetime is 'like a gigantic jelly fish that can be squeezed and bent', within which we are immersed. These, I believe, are not trivial metaphors for popular science: this is powerful scientific imagining, which has propelled and nourished our best science.

Advancements in science often amount to discoveries of a new perspective, a literal 'change of point of view'. But modern physics is also revealing a deep underlying perspectival structure of reality. In relativity, gauge theory, quantum field theory when gravity is not neglected, probably in quantum theory, and perhaps even in the flow of time, we understand phenomena relationally: in terms of how systems appear from the perspective of other systems.

To describe the world, we need to take this deep relational, perspectival, structure of the world into account. Not only because we access the world always only from a specific perspective, but also because the world itself is probably better described by relating perspectives than by attempting a perspective-free view 'from nowhere'.

References

[1] Einstein, A. *Autobiographical Notes.* Chicago, IL: Open Court, 1991.

[2] Rovelli, C. *Anaximander.* Chicago, IL: Westholme, 2011.

[3] Popper, K. *The Logic of Scientific Discovery.* London: Hutchinson, 1959.

[4] Kuhn, T. *The Structure of Scientific Revolutions.* Chicago, IL: University of Chicago Press, 1962.

[5] Finetti, B. de. *L'invenzione della verità.* Milan: Raffaello Cortina, 2006.

[6] Ramsey, F. *The Foundations of Mathematics and Other Logical Essays.* New York, NY: Harcourt, Brace and Company, 1931.

[7] Clark, A. Whatever next? Predictive brains, situated agents, and the future of cognitive science. *Behav. Brain Sci.* 2013; **36**(3): 1–73.

[8] Taine, H. *On Intelligence,* trans. T. D. Haye. New York, NY: Henry Holt and Company, 1889. https://archive.org/details/onintelligence01taingoog/page/n4/mode/2up.

[9] Quine, W. *Ontological Relativity and Other Essays.* New York, NY: Columbia University Press, 1969.

[10] Bogdanov, A. *Tektologiya: Vseobschaya organizatsionnaya nauka.* Berlin, Petrograd, and Moscow, 1922; *Essays in Tektology: The General Science of Organization,* trans. G. Gorelik. Seaside, CA: Intersystems Publications, 1980.

[11] Wiener, N. *Cybernetics; or, Control and Communication in the Animal and the Machine.* Paris: Technology Press, 1948.

[12] Bertalanffy, Ludwig von. *General System Theory: Foundations, Development, Applications.* New York, NY: George Braziller, 1968.

[13] Ismael, J. Do you see space? How to recover the visible and tangible reality of space (without space). In Wüthrich, C., Le Bihan, B., and Huggett, N., eds. *Philosophy Beyond Spacetime.* Oxford: Oxford University Press, forthcoming.

[14] The Event Horizon Telescope Collaboration. First M87 Event Horizon Telescope results. I. The shadow of the supermassive black hole. *Astrophys. J. Lett.* 2019; **875:** L1.

[15] Unruh, W. G. Notes on black-hole evaporation. *Phys. Rev. D* 1976; **14**(4): 870–892.

[16] Rovelli, C. Why gauge? *Foundations Phys.* 2014; **44:** 91–104. https://arxiv.org/abs/1308.5599

[17] Vaidman, L. Many-worlds interpretation of quantum mechanics. In Zalta, E. N., ed. *The Stanford Encyclopedia of Philosophy* (Fall 2018 edition). https://plato.stanford.edu/archives/fall2018/entries/qm-manyworlds/

[18] Laudisa, F., and Rovelli, C. Relational quantum mechanics. In Zalta, E. N., ed. *The Stanford Encyclopedia of Philosophy* (Summer 2013 edition). https://plato.stanford.edu/archives/sum2013/entries/qm-relational/

[19] Rovelli, C. Is time's arrow perspectival? In Chamcham, K., Silk, J., Barrow, J. D., and Saunders, S., eds. *The Philosophy of Cosmology.* Cambridge: Cambridge University Press, 2016; 285–296. http://philsci-archive.pitt.edu/11443/

5 Vision of the Cosmos

CAROLIN CRAWFORD

Astronomy is the science of testing how the fundamental rules of physics operate in extreme conditions – extremes of distance, time, and scale, and of physical properties such as temperature, pressure, density, and mass. Such conditions are not accessible in the lab, so the traditional scientific methodology employing a 'control' and an 'experiment' is not available. All we can learn about the wider cosmos beyond the Solar System depends on our ability to capture, analyse, and interpret the photons of light that happen to fall towards Earth; and how that light has been radiated, reflected, absorbed, and refracted along its path.

Early observers were dependent solely on the unaided eye, which can pick up the light from about 6,000 stars, ranging over a factor of 100 in brightness. They also monitored the movements of the Sun, the Moon, the planets out as far as Saturn, and any comet against the fixed backdrop of constellations. Not surprisingly, ancient cultures developed a view of the cosmos that revolved around us: both figuratively, in that the motions of objects in the heavens were interpreted as astrological signs and omens bringing messages to humanity; and literally, where the understanding was that everything in the cosmos was in a physical orbit centred on the Earth.

This view changed radically once Galileo turned a telescope to the heavens in 1609, providing the first new astronomical data for thousands of years. Even through the crudest of instruments, his observations of the phases of Venus and the movements of Jupiter's moons were ammunition for the removal of Earth's special place at the centre of the Heavens. But, importantly, Galileo also demonstrated that there were stars in the night sky that were invisible to the unaided eye, and thus were not known to exist unless you viewed them through a telescope. A further step change in the way we carry out observations came when the British astronomer Sir William Herschel discovered infrared radiation in 1800, showing that there weren't just *objects* invisible to the unaided eye, but also forms of *light*. Modern astronomers observe in all these invisible colours of the electromagnetic spectrum – from lower-energy photons in the infrared, sub-millimetre and radio waves emitted from cool regions of space, to the high-energy ultraviolet, X-rays, and gamma-rays radiated by the hottest and most violent events in the Universe.

During the 400 years of observations that separate us from Galileo, humanity's perspective of our place in the cosmos has completely changed. The Heavens are no longer centred on Earthly concerns; we now know our home is only one among millions of planets around billions of stars in our Galaxy, which in turn is only one of hundreds of billions of galaxies. One of the most important things we've learnt is that what we observe in any waveband is only 5 per cent of what is out there. The remaining 95 per cent of the matter–energy content of the Universe is comprised of mysterious dark matter and dark energy – vast swathes of the cosmos that are apparent only by the way they pull or push observable matter around. Much of our knowledge has been made possible by the continuing development of new facilities, both on the ground and on satellites out in space, thereby opening up new wavebands to observation, and improvements in the design of telescope and detectors. Observations have driven much of the science over the last 400 years, and it is reasonable to assume that they will continue to do so into the future. I shall highlight some of the amazing observational facilities that will come online over the next decade or two, discussing how they'll build on and extend previous discoveries – and how they will fundamentally change our way of doing astronomy.

In practice, the process of observation hasn't changed so much from that used by an astronomer/astrologer from an ancient culture. We still gather light and analyse the data, only now we use a telescope rather than our eyes, and computers assist the brainwork involved. What we demand from any telescope is much the same as we would wish from our eyes: we want to gather as much light as we can; in as many different colours as possible; we want our images to be in focus, over a large field of view; to resolve detailed structure; and be sensitive to changes with time.

We need to gather as much light as possible because objects in the Universe are faint. The further away something is, the more its light is diluted by the inverse square law, and the dimmer it appears to us. Astronomers are viewing cosmic objects over such immense distances that the finite travel time of light matters; sometimes we are collecting light that left its origin millions to billions of years ago. The light thus brings information about that source as it was millions to billions of years ago – not only are further objects fainter, but we're seeing them as they were further back in the past. Also, in any population the faint objects are far more numerous than the bright ones. If we want to make statistical comparisons within a population at any epoch, we need to study a *representative* population rather than only catch the light from the brightest, and potentially more exceptional, examples.

Large-Aperture Telescopes

The basic premise of improving a telescope's ability to collect as much light as it can is that *bigger is better*. The wider the collecting area of the telescope (i.e. the diameter of the main mirror/lens), the more photons it will catch, and the brighter the source will appear. The power of a telescope is routinely described by its *aperture*, as this determines how faint the objects it can observe can be. The very largest telescopes in operation today use mirrors 8–10 m across, enabling them to collect 4 million times more light than a human eye: examples include the Very Large Telescope, the Gemini Telescopes, and Subaru, all 8 m in diameter; and the two Keck telescopes, each 10 m across. The next generation of ground-based telescopes will have apertures of up to 40 m across – accompanied by suitably astronomical budgets (over $1,000 million). The larger mirrors are constructed from tiled hexagonal segments, each a sizeable mirror in its own right. Currently there are three 'extremely large' telescopes in the offing. The Giant Magellan Telescope (GMT) will have a 24.5-m aperture comprised of an arrangement of seven 8.4-m mirrors; this is currently under construction in Chile, and expected to be ready for use in 2025. The Thirty Meter Telescope (TMT) in Hawaii will have a 30-m mirror constructed from 492 hexagons, each 1.4 m across; construction has been approved, with expected completion

FIGURE 5.1 An artist's impression of the ESO's Extremely Large
Telescope with a tiled 39-m diameter mirror. Credit: ESO/L. Calçada.

in about 2027. Finally, the Extremely Large Telescope (ELT) will be
39.3 m across, composed of 798 1.4-m-diameter segments. It is already
under construction in the Atacama Desert in Chile, and expected to be in
operation in about 2024 (Figure 5.1). This telescope alone will provide
over 250 times the light gathering area of the Hubble Space Telescope,
and provide images 16 times sharper.

One of the major goals for the next generation of telescopes is to study
the most distant galaxies. The deepest views of the Universe to date are
obtained when a telescope stares at one tiny patch of sky for a prolonged
period, allowing us to gather light from galaxies at a range of distance
away from us in one field of view. These data reveal how galaxies evolve
and change with time. Comparing galaxies from different epochs shows
that those seen as they were a few billion years ago display a much more
irregular morphology than the galaxies we see around us in the current
Universe – often displaying a blobby structure, both spread out and

elongated into chains. The bluer colours imply that significant star formation is taking place. We interpret the evolution of these types of galaxies to be hierarchical, where smaller proto-galactic fragments slowly assemble and merge together to form larger structures. There is still the concern that we are picking up only the brightest, largest, and most dramatic examples at any epoch, though other methods can reveal random background more serendipitously, for example through the magnification provided by the gravitational lensing effect of a foreground cluster of galaxies.

The number of truly distant galaxies known is small. The very earliest we can observe date back to when the Universe was about 400–500 million years old. Spanning a few thousand light-years across and weighing only a few million solar masses, they are about 1/100th the mass and size of a fully grown galaxy like the Milky Way. Fewer stars means these galaxies are inherently less luminous, even without the dilution of this luminosity by their extreme distance away from us. At such large distance away from us, most of the visible starlight is redshifted away from the visible and into the red and infrared.

It's not enough to simply discover distant galaxies in sufficient quantities, rather we need to be able to resolve the detailed structure. Only with enough photons from each component of a blobby image will we be able to determine the colours and spectral information that reveal the physical processes – such as where star formation is occurring. We express the ability of a telescope to see detail in terms of its *angular resolution*, which is the minimum angle that separates the appearance of two stars still seen as distinct sources. The smaller the angular resolution, the finer the details that can be made out in an image.

The angular resolution of a telescope is limited by diffraction, which is the tendency of light waves to spread out from their point of origin. There is an absolute limit that a telescope can achieve, given by the equation $\theta = 1.22\lambda/D$, where the wavelength of light (λ) and the diameter of the aperture (D) are related to give the angular resolution θ in radians; the smaller the angle θ obtained, the better the resolution. Thus a wider aperture gives less diffraction and sharper detail at any wavelength. This is only a *theoretical* best, however, as even the largest-aperture ground-based telescopes have their angular resolution compromised by the fact that the light they collect first has to travel through our atmosphere. Pockets of micro-turbulence in the air cause continual refraction of the light, making it seem as if the images dance around. This 'twinkle' of starlight is unpredictable, and changes on time scales of less than a second, smearing light collected over time into a blob rather than the pin-point it should be. Ground-based telescopes attempt to sidestep the distortion of images by using complex adaptive-optics systems where the deformation of nearby artificial stars (generated by lasers beamed from the telescope into the sodium layer of our atmosphere) is continually monitored so that adjustments can be made to the mirror shape in real time to correct for the atmospheric turbulence.

Space Telescopes

The better – but more expensive! – way to avoid atmospheric blurring altogether is to observe from space. The Hubble Space Telescope (HST) is such a fantastic success not because it has a tremendously large mirror (it is only 2.4 m in diameter), but because it was collecting light above the atmosphere. Hubble was launched in 1990 and is now reaching the end of its working life, to be succeeded by the James Webb Space Telescope (JWST), due for launch in March 2021. The JWST's 6.5-m-wide primary mirror consists of 18 hexagonal mirror segments made of gold-coated beryllium, and it will bring 100 times the sensitivity and 10 times the image sharpness of the HST (Figure 5.2). The mirror has to pack away into the nose cone of a rocket for launch, and will unfold to its full diameter only once it is in space. The JWST will not operate in low Earth orbit like the HST, but will observe from a vantage point some 1.5 million kilometres beyond Earth's orbit, near the Earth–Sun L_2 (Lagrange) point, where a single large sunshield can be deployed to block the heat and light both of the Sun and of the Earth.

A sunshield is required in order to keep the temperature of the spacecraft below $-220\,°C$. Unlike the HST, which worked in a broad range of light from the near-ultraviolet through to the near-infrared, the JWST will access only the longer-wavelength bands from the red to the mid-infrared range. It is thus vital to keep the telescope and its spacecraft so cold to prevent it radiating in the very infrared colours it's trying to

FIGURE 5.2 The primary mirror of the NASA JWST showing its design from 18 hexagonal mirrors. Credit: NASA/Chris Gunn.

detect, which would otherwise produce an annoying background glow. Operating in the infrared is also a crucial advantage (along with the light-gathering power and the high angular resolution) for detecting distant objects. This is because, while light is travelling across a volume of space, it is stretched with that space by the expansion of the Universe. Lengthening the wavelength reddens the light, and thus light from the most remote sources moves entirely into the infrared waveband.

A key ambition for the JWST and future telescopes is to look further back than the current limit of a few hundred million years after the Big Bang. While the most distant galaxies we can observe may be early versions of galaxies much like our own, they are still galaxies; astronomers want to see when and how galaxies *start* forming, a regime that currently lies beyond the reach of current ground- and space-based instruments. The Nancy Grace Roman Space Telescope (NGRST) is another infrared space observatory that is being developed for launch in the mid 2020s, which will observe from the Earth–Sun L_2 point. It will have a smaller mirror than the JWST, of only 2.4 m in diameter, but its field of view will be over 100 times larger. This wide field allows the NGRST to survey much of the sky, and discover objects that can then be more comprehensibly characterised in follow-up observations with the JWST. The lead time to develop a major new facility is so long that NASA is already planning for a successor to the JWST and the NGRST. The idea is that they will be superseded in the 2040s by a multi-wavelength space observatory called the Large UV Optical Infrared Surveyor (LUVOIR). Designed to work in the ultraviolet, visible, and near-infrared wavelengths of light, it will probably have either an 8-m or a 15-m mirror.

Better sensitivity and angular resolution will not only reap dividends for observation of the very distant Universe, but also allow us to see nearby objects in our own galaxy in more detail. Both the JWST and the NGRST will study *exoplanets* – planets in orbit around stars other than the Sun. We know of over 4,000 exoplanets, and perhaps another 20,000 are expected to be discovered by the Transit Exoplanet Survey Satellite (TESS) launched in 2018. With a more complete census of exoplanets, we will be able to answer questions about the potential for life in the Universe, and how representative the Solar System is of other planetary systems in a wider context.

But we are now beyond simply discovering and cataloguing the types of exoplanets out there. The large field of the NGRST will find exoplanets down to a mass only a few times that of the Moon through gravitational microlensing, and both the NGRST and the JWST will isolate the light of larger planets. Such images will resolve an exoplanet only as a tiny dot, but can trace variations both in its colour and in its light as it revolves along its orbit. Follow-up observations of planets with Earth-like masses that lie at close separation from their host will yield detailed characterisation of their atmospheres. This is science that we already do with the HST, but it's very expensive in telescope time and so far has been carried out only for a handful of the larger exoplanets. The technique uses spectroscopy – measuring the intensity of light at different wavelengths – of the light from the host star when its planet has moved between us to pass in front of the star. The way the starlight is filtered through the planet's atmosphere reveals the gases it contains. Different elements and molecules absorb light at specific energies to produce absorption lines that are a tell-tale signature of their presence, allowing us to characterise the composition of a planet's atmosphere. The combination of the imaging of reflected light and spectroscopy of filtered light can also potentially tell us about the weather on the planet – whether clouds at high altitude blanket the lower levels, or the temperature profile within the atmosphere which could constrain circulation models for the transport of heat between day and night sides. Molecules in the atmospheres of exoplanets are likely to have the largest number of spectral features in the infrared waveband, and the JWST is required to extend the search for signs of water vapour and other molecules to smaller-mass (super-Earths and Earth-like) planets – as well as potential bio-signatures that could trace the presence of life.

The JWST also has the potential to revolutionise our understanding of how stars and planets come into being. Star formation is the mechanism that underpins the structure of galaxies, so it is essential to deconstructing their evolution through time. Star formation occurs within interstellar space where it is at its most cold and dense, and the pull of gravity is able to overwhelm any outwards thermal pressure. Stars form deep within cocoons of dusty opaque gas, completely hidden from (visible) view until the newly fledged star clears its surroundings to burst into view as a *fait accompli*. Our understanding of the star-formation process comes more from computer modelling than from direct observation. The problem is exacerbated for very massive stars; not only are they rarer – meaning that there are few nearby examples – but also they involve more mass and so are more heavily obscured. Their gravitational collapse into stars happens very rapidly as the force of gravity pulling matter together is much stronger. Longer-wavelength light such as at near-infrared, sub-millimetre, and millimetre wavelengths can penetrate this fog, since the photons are less likely to be scattered away from our line of sight by the dust particles as they try to escape from the centre of the otherwise opaque cloud. This leads to the possibility of watching star formation for the first time and observationally understanding what factors (temperature, pressure, density, magnetic field, . . .) are important for tipping a cloud to the point of gravitational collapse. What determines whether a cloud will form a single star or a binary system, or influences the mass of stars that condense from a cloud?

Once a young protostar has started shining, it is still surrounded by a thick cocoon of dusty material, known as a *proto-planetary disc*, where a planetary system is slowly forming. The HST first detected these discs around young stars by their obscuration of background light; but the longer wavebands and higher resolution of the JWST will permit us to see inside to witness the process of planetary formation. Within these discs, neighbouring dust particles start to stick together by electrostatic forces, slowly growing into grains, which then accumulate to the size of small pebbles and larger-sized rocks. These collide and adhere to build proto-planetary bodies, eventually acquiring enough mass that their gravity can dramatically reshape the structure of material in the surrounding thin disc. As planets form, they sweep their orbits clear of debris to create gaps in the disc, and their gravity sweeps the dust and gas up into tighter and more confined rings. We can already see this process, in observations of young stars (less than a million years old) taken with the ALMA telescope (see the next section) in the sub-millimetre waveband. Gaps and asymmetries revealed in the distribution of dust in the disc mark where large planets are accumulating matter. The structures can be matched to detailed computer simulations modelling the coalescence of material in the disc to form planets. A wider survey of this process around stars of a range of ages should give us the time scale for formation of a planetary system, the chemical and physical properties of the disc, including the distribution of volatile molecules such as water, the ratio of gas to dust in the disc, and how this influences the kind of planets that form.

Interferometry

The Atacama Large Millimeter/sub-millimeter Array (ALMA) Observatory is located in the dry Atacama Desert of northern Chile, and operates beyond the infrared, in the millimetre and sub-millimetre wavebands. At such wavelengths, a dish antenna rather than a mirror is necessary to collect the light, but the principle that a wider aperture is

FIGURE 5.3 ALMA sub-millimetre observation (in false colours) of the structure within the planet-forming disc around the young star HL Tau. The concentric rings show where forming planets are sweeping their orbits clear of debris, and sweeping dust and gas into confined zones. Credit: ALMA (ESO/NAOJ/NRAO), NSF.

better still applies. The angular resolution is worse than in the visible wavebands; recalling the equation $\theta = 1.22\lambda/D$, one sees that, as the wavelength has now increased, the achievement of a comparable image resolution would require a dish too large to build. An ingenious way to get round this is to combine radio dishes in an array that can mimic the operation of a much larger dish, in a technique known as *interferometry*, which has been standard practice in radio astronomy for many years. With careful coordination of the observations, higher-resolution images can be obtained (Figure 5.3).

ALMA uses 66 dishes of diameter 7 m and 12 m, which are mobile and can be transported across the desert to separations between 150 m and 16 km. The distance between dishes determines the balance between the resolution you want to achieve (wider is better, so the best resolution is similar to that from a single dish of diameter 16 km) and over how large a source (a more compact array gives a larger field of view). The longer wavelengths mean that radio interferometry is easier than in the optical, but the synchronisation of dishes for pointing and capturing an identical signal from the target and the subsequent correlation of the collected data have to be carried out with a precision better than a trillionth of a second. ALMA has been fully operational since March 2013 and has achieved a range of new results, including measurements of the distribution and complexity of molecules in the interstellar clouds that are the raw material for star formation.

The principle of interferometry is being continued with two very ambitious projects. The Square Kilometre Array (SKA) telescope is an international project to construct the world's largest radio telescope, using thousands of dishes up to 15 m across, and up to a million smaller antennae, to sample a total collecting area of over 1 km^2. It will operate in low-frequency (i.e. long-wavelength) radio bands, producing radio images at comparable resolution to the HST in a wide range of radio colours, and with a sensitivity 50 times better than any other radio facilities. Such a telescope has to be located in radio-quiet areas, and construction will be split between two remote sites in South Africa and Australia. The project is undergoing testing and design work ready for the first phase of construction due to start in 2024, with the first trial observations expected by 2028 (Figure 5.4).

The SKA will be used to survey the whole sky, and in particular will map one of the lesser-known components of our Milky Way, the diffuse cold gas that comprises the interstellar medium. The interstellar gas which is detected in both neutral atomic and molecular form represents

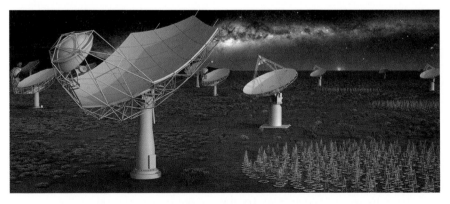

FIGURE 5.4 An artist's impression of the SKA in operation, featuring both the dishes and smaller antennae. Credit: SKA Organisation.

the reservoir for star formation, long before the process of gravitational collapse begins. Thus the observed distribution of molecular gas helps tell us where, and how fast, galaxies form new stars – not just in the spiral arms of our Milky Way, but also in more remote galaxies where we can see how the properties and processes influencing star formation might differ from one galaxy to another.

The SKA is intended to resolve a major issue about star formation much further from home, since it has the sensitivity to probe a period in the Universe's early history known as the 'dark ages', which so far has remained inaccessible to observation. By the time the Universe was about 300,000 years old, it had expanded and cooled down sufficiently that the very first atoms of hydrogen were able to form. This process of *recombination* moves the Universe from being opaque to transparent; but, while photons of light can travel through it towards us, it remains dark because there aren't any stars or galaxies yet to radiate light. After recombination, space is simply filled with neutral hydrogen, emitting faint radio emission that should be detectable with the SKA. At some point during these dark ages, the very first generation of stars will begin to condense from this primordial hydrogen gas. They are likely to be incredibly massive – each 100 times or more the mass of the Sun – and thus very short-lived. Their transitory nature and remoteness means that we have no hope of seeing individual stars themselves, but we can observe the impact they have on the early intergalactic medium. Being hot and luminous, such massive stars give out intense ultraviolet radiation with enough energy to break apart the atoms in the gas around them. As electrons are stripped from the hydrogen, the surrounding neutral atomic gas steadily changes into an electrically charged plasma. Early luminous objects – the very massive stars, followed by the first baby galaxies – produce increasing levels of radiation that break apart the neutral hydrogen in a growing bubble around them. The bubbles gradually overlap, changing the neutral intergalactic medium into one that is almost completely reionised; this is the epoch of *reionisation*. The SKA will trace the evolution of the neutral/ionised properties of the gas in the early Universe as an indirect way to date the emergence of the very first structures in the Universe, giving us insight into a crucial period in cosmic history.

Carolin Crawford

The Event Horizon Telescope (EHT) uses interferometry to construct an array with the diameter of the whole globe.

Linking up eight radio dishes, including ones at the South Pole, the Andes, Hawaii, and Europe, at millimetre wavelengths mimics an aperture with an angular resolution capable of isolating something the size of a grapefruit on the surface of the Moon. This is sufficient to resolve the event horizon of two of the closest supermassive black holes to Earth: Sgr A*, 26,000 light-years away at the centre of our Milky Way, with a mass 4 million times that of our Sun; and the 9.5-billion-solar-mass behemoth at the core of the giant elliptical galaxy M87 55 million light-years distant. The EHT spent 10 days continually observing these supermassive black holes in 2017–2018. The correlation and processing of all the disparate radio signals was a complex process that took months to do, as precise timing is required in order to account for the different arrival times of the radio waves. Note that the resolution does not match the black hole itself (which is the singularity), but instead the point of 'no return' around it known as the *event horizon*; this is the boundary beyond which you cannot see any events because light cannot escape from within it. The images of the black hole in M87 (those of Sgr A* will take much longer to produce) reveal the distorted shadow of the event horizon against the background glow of hot gas being accreted by the strong gravitational field. This is the first observable confirmation that the idea of a black hole having an event horizon is correct, and the appearance of the image can be tested against quantitative predictions expected from whether or not the black hole is spinning.

X-Ray Astronomy

Observations of black holes of all sizes (i.e. both stellar-sized and the supermassive kind) are most readily done in the X-ray waveband. We don't observe the black hole itself, as light can't escape the extreme gravity, but, because X-rays are the most energetic light, they are emitted by accreting matter just outside the event horizon. X-ray observations thus sample the extreme distortion of space and time in a strong gravitational field, allowing us to test our understanding of physics, and in particular of General Relativity under the most extreme conditions.

The accretion process is responsible for launching hot winds and energetic jets that blast out from supermassive black holes, capable of carrying out energy that is later dumped into the wider surrounding galaxy. This heating can impact on the rate of star formation on large scales, limiting the growth of the surrounding host galaxy – and thereby eventually starving the black hole itself of fresh fuel, regulating its own mass in turn!

Various small missions are planned for the next few years, but the step change for X-ray astrophysics will be provided by the Advanced Telescope for High ENergy Astrophysics (Athena) (Figure 5.5), which is currently under development by the ESA. Athena will be 100 times more sensitive than current X-ray telescopes. Since the Earth's atmosphere blocks most X-ray emission, Athena will be a space telescope, and is due for launch in about 2031. Though the design of an X-ray telescope is very different from the radio dishes/optical mirrors (in that the mirrors have to be presented almost edge-on to the source of light rather than face-on), the principle is still to provide an X-ray mirror with a very large collecting area, and sensitive instruments capable of

© ESA, IRAP, CNES, XMM-Newton & ACO.

FIGURE 5.5 An artist's impression of the Athena X-ray satellite looking at the centre of the Milky Way. Credit: ESA/IRAP/CNRS/UT3/CNES/Fab&Fab.

detecting photons in a range of energies at its focus. The improved capacity provided by Athena will enable us to track the activity of supermassive black holes out to the highest redshifts, out to the time when star formation in galaxies and accretion processes at their centres were most intense. Given that the feedback from central supermassive black holes most probably plays an important role in determining the evolution of galaxies, more distant examples should help us track how massive galaxies develop.

Black holes are fundamentally powered by unstable flows of matter falling under gravity, so the luminosity output depends on fluctuations within the fuel supply. At high energies, accreting sources appear to burst, flicker, and pulsate in brightness, on time scales from hours and days to weeks and months. X-rays are also generated in the most energetic/ explosive events in the Universe, such as when a massive star undergoes an enormous outburst at the end of its life to produce a supernova. A crucial feature of Athena is its very rapid response time, enabling it to slew to catch a source freshly in outburst, or a supernova event.

Time-Domain Astronomy

The wish to respond rapidly to observed events is an important illustration of one way in which astronomical observation is changing. Although most cosmic objects live and evolve on time scales of millions or billions of years, certain types of astronomical source produce changes in their colour, brightness, shape, spectrum, and position. Variations can occur on time scales of less than seconds to weeks, month, and years – or in far more dramatic one-off *transient* events. Although the large-aperture telescopes we have discussed enable us to see further and fainter, they can be used only to look at selected narrow regions of the sky, targeting individual sources. The new era of *time-domain* astronomy involves a rapid and continual survey of objects in the sky, to map not only the locations of cosmic sources but also how they change with time. We use automated telescopes to do this, each with a large field of view that enables it to observe a large part of the sky at a time. Powerful computers undertake routine but rapid analysis and comparison of the data, rapidly flagging sudden changes as possible targets of interest to ensure detailed follow-up observation of an event with other telescopes operating in a variety of wavebands.

The most exciting prospect immediately on the horizon is the Large Synoptic Survey Telescope (LSST), a survey telescope with an 8.4-m primary mirror that is currently under construction and due to begin operation in 2020.

Not only will it have very sensitive detectors able to pick up faint objects in short exposures, but also, by using the largest digital camera yet built, it will have an exceptionally wide field of view. Able to encompass a patch of the sky seven Moon-widths on a side, a single image from the LSST will be equivalent to 3,000 images from the HST. This enables the LSST to image the entire sky about twice a week, taking over 200,000 pictures a year. Extracting the scientific outcomes from the amount of data collected – particularly if the intention is to produce rapid alerts – will require considerable advances in automated data analysis.

Close to home, such surveys should discover the population of dark asteroids that present a potential future threat if their path crosses the Earth's orbit, and many more objects such as Oamuamua, thought to be interstellar interlopers – fragments of rock torn from distant planetary systems to temporarily plunge through the centre of the Solar System. But the LSST is particularly expected to revolutionise the discovery of supernovae, by locating over 100,000 a year. Observing supernovae doesn't just allow us to test our understanding of the demise of stars at the end of their life; observations of a particular kind of 'Type 1A' supernova (caused by transfer of mass in a binary system) are used to measure the geometry of the Universe, offering the possibility of constraining the behaviour of dark energy. The discovery that remote Type 1a supernovae were observed to be fainter – and thus more distant – than indicated by their redshift presented the first evidence for the accelerated expansion of the Universe, and indicated that the matter–energy content of the cosmos is predominantly dark energy rather than in the form of matter. The trouble is that such supernovae are rare events, with on average one exploding per century per galaxy. Up to now we've had to rely on the chance luck of catching one in a far-off galaxy, but survey telescopes will offer a systematic search, catching them rapidly enough that we get the clearest information at the peak of the outburst. Using further and fainter supernovae as tracer particles will allow us to map out the expansion of the Universe to far greater distance, and where the rate of change in the accelerated expansion of the Universe might provide a discriminant between theories competing to account for dark energy.

Gravitational-Wave Observation

A stunning development in recent years has brought astronomy into a new phase where we no longer have to rely on just light for our observations, but can now detect gravity directly by sensing the minuscule squeezing and stretching distortions of space produced when the ripple of a strong gravitational wave passes through Earth. A gravitational wave represents a message about the change in the gravitational field spreading out into the Universe. Unlike the electromagnetic waves, which travel *through* spacetime, gravitational waves are travelling distortions in the shape of space itself. So far 10 cases of merging black holes have been detected from ground-based detectors such as those of the Laser Interferometer Gravitational-Wave Observatory (LIGO). Interestingly, gravitational waves are sensitive to a hitherto undiscovered flavour of black hole; still stellar-sized, but of larger mass (over 20 solar masses) than those regularly observed through X-ray telescopes. A sensitivity upgrade for LIGO will soon allow it to sense black-hole collisions almost twice as far away, and by 2024 it is expected to find more than three black-hole merger events a day, as well as more mergers of two neutron stars. Such a sample will provide better statistics about the frequency of such events, the kind of progenitor black holes involved, their mass, and their spin.

Ideally, we would like to detect gravitational waves from an even more dramatic collision, such as that of the supermassive black holes at the cores of two large merging galaxies. The typical time scale for stellar black-hole mergers is milliseconds, but supermassive black holes would merge over about 1,000 seconds. To collect gravitational waves at these low frequencies requires a detector a million times longer than the 4-km-long laser beams used in ground-based experiments such as LIGO; thus the future of gravitational-wave detection will be out in space. One such experiment planned is the Laser Interferometer Space Antenna (LISA), consisting of three spacecraft arranged in an equilateral triangle with sides 2.5 million kilometres long, which will trail about 50 million kilometres behind the Earth in its orbit around the Sun. The distance between the satellites will be precisely monitored to detect a passing gravitational wave, which would alternately squeeze and stretch their separations by a tiny amount. Even given the intended launch date of 2034, this presents a huge technological challenge.

In 2017 not only was a collision and merger of two neutron stars detected by three separate gravitational-wave experiments, but also the event was followed 1.7 seconds later by a burst of gamma-rays seen by two satellites. None of these detectors had good resolution by itself, but triangulation of the results narrowed the source location to an area of sky about 60 Moon-widths on a side. The observation triggered an automatic alert to other telescopes, and soon a team at the Los Cumbres Observatory in Chile discovered a bright new source in the visible, located in a galaxy some 130 million light-years from Earth. Follow-up observations over the next few weeks revealed the source to be light from a long-hypothesised *kilonova*. A kilonova occurs when a small amount of material left over from the collision of two neutron stars is ejected at speeds as high as 30 per cent of the speed of light. Not only did the discovery confirm that such a process could be responsible for at least some of the mysterious short bursts of gamma-ray light, but also it was the first time that any cosmic event had been observed both in gravitational waves *and* in light.

We learn complementary information from each observation: the gravitational waves reveal the motions and masses of the neutron stars involved in the merger, while the electromagnetic radiation reveals the astrophysical processes underpinning the event. Such *multi-messenger astronomy* is the future, where we try to combine all available information about a cosmic source. And there are two further 'messengers' that astronomers are beginning to employ.

Cosmic-Ray Observation

Despite their name, cosmic rays are not beams of light, but energetic particles that arrive from all directions in space to bombard the top of

Earth's atmosphere. They are electrically charged fragments of atoms that move at speeds close to that of light and are the most energetic particles we can observe. Cosmic rays provide one of our few direct samples of matter from outside the Solar System, and somewhere between 10 and 100 fly through your body every second.

You can't see cosmic rays, and those with the very highest energy can't be measured directly; instead we detect that they have interacted with the nuclei of atoms in our upper atmosphere, to produce a cascade of lower-energy secondary particles which can be measured by detectors on the ground. Simulations are used to reconstruct the properties of the primary particle. The current state of the art for high-energy cosmic-ray detection is the Pierre Auger Telescope, an international facility located in Argentina with detectors spanning a collection area of 3,000 km^2, which was completed in 2008. Cosmic rays cannot be used to reconstruct an image of their source, as we can't track them back to their origin – because they are electrically charged, their path has been influenced by any magnetic fields en route. But the characterisation of the properties of detected cosmic rays does allow us to make inferences about various physical processes, occurring within the Galaxy and beyond, capable of accelerating such particles to nearly the speed of light.

Neutrino Astronomy

The last piece of the jigsaw is *neutrino astronomy*. Neutrinos are energetic particles that flood the Universe at an estimated density of over 100 per cubic centimetre. They have barely any mass and are produced from nuclear reactions such as fusion, or radioactive decay. Neutrinos are difficult to catch because they interact only very weakly with matter – most that intercept the Earth just pass directly through. While this lack of interaction makes them incredibly hard to detect, it is also what makes them so exciting, as it means that they can escape the incredibly dense environments in which they are produced. If we could detect them directly, they would give us a completely different view of the Universe: light provides information only about the surface of cosmic objects, whereas neutrino astronomy provides a way to trace processes occurring at their heart. This could be processes at the core of stars, or the final implosion of matter in a supernova; or we could observe neutrinos created at the time of the Big Bang or even as a by-product of the decay or destruction of hypothesised particles of dark matter.

For a long while neutrino astronomy was limited to the detection of neutrinos both from the Sun and from the 1987 supernova that exploded in 1987 in a satellite galaxy to the Milky Way. Higher-energy neutrinos are more exciting as they are expected to be generated by the most violent astrophysical sources: events like exploding stars, gamma-ray bursts, and those involving black holes and neutron stars. Some of the likely processes are expected to generate both high-energy neutrinos *and* high-energy gamma-rays, so linking the neutrino and high-energy gamma-ray detections is useful. High-energy neutrinos can currently be detected only at the IceCube Observatory at the South Pole, where 86 strings, each strung with 60 sensitive light detectors, have been sunk deep into a cubic kilometre of water ice (Figure 5.6). These thousands of light detectors continually search the ice for the brief flash of light revealing that a passing neutrino has collided with a nucleus of an atom, giving a minute injection of energy. The direction of the neutrino's arrival can be determined by the flashes tracing its path through the ice. As neutrinos are electrically neutral, unlike cosmic rays they are unaffected by any electromagnetic fields as they travel through space, so with enough detections there is the potential of being able to trace a direct path back to their origin on the sky. IceCube has already detected

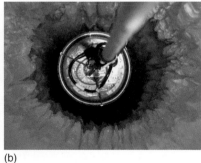

(a) (b)

FIGURE 5.6 (a) The IceCube Observatory in Antarctica and (b) one of the strings of sensitive light detectors being deployed into the ice. Credit: Felipe Pedreros (a) and Mark Krasberg (b) IceCube/NSF.

over 100 exceptionally fast-moving neutrinos, including some with over 100 times more energy than we can accelerate particles to in the Large Hadron Collider. The success of IceCube has paved the way for an expansion of the current facility into a more efficient detector. The ice is sufficiently transparent that adding an array of more light sensors but at a wider spacing will allow the search for neutrino collisions to take place over a volume of 10 km^3, potentially yielding an improvement in detection rates of an order of magnitude. This would move us into a regime of collecting statistically significant samples of very-high-energy astrophysical neutrinos, enabling detailed spectral studies and determination of sources.

In many ways there has never been a more exciting time in astronomy, where all our ideas – from the nature of dark energy driving the accelerated expansion of the Universe down to the architecture of planetary systems – will develop as the new data roll in. This chapter has focused on the definite science objectives and ambitions that have shaped the specifications and design of all these new telescopes and missions. We should remember, however, that the ultimate return from all these facilities will include new investigations and discoveries that we cannot begin to imagine today.

Further Reading

The design and building of astronomical telescopes is a fast-evolving field, so the webpages of individual facilities (listed in order of their mentions in the text of this chapter) provide the most up-to-date information on implementation and science objectives.

GMT	www.gmto.org
TMT	www.tmt.org
ELT	www.eso.org/public/unitedkingdom/teles-instr/ elt
JWST	www.jwst.nasa.gov
NGRST	www.nasa.gov/content/goddard/nancy-grace-roman-space-telescope
LUVOIR	https://asd.gsfc.nasa.gov/luvoir/

TESS	www.nasa.gov/tess-transiting-exoplanet-survey-satellite
ALMA	www.almaobservatory.org/en/home
SKA	www.skatelescope.org
EHT	www.eventhorizontelescope.org
Athena	www.the-athena-x-ray-observatory.eu
LSST	www.lsst.org
LIGO	www.ligo.caltech.edu
LISA	www.lisa.nasa.gov
Pierre Auger Observatory	www.auger.org
IceCube	www.icecube.wisc.edu

6 Visions of a Digital Future

SOPHIE HACKFORD

I am going to explore how machines see us, and how we see them.

Machines are starting to view us in exquisite detail, which heralds the start of a visual conversation between us and the algorithms. The more they see of us and the world, the better this conversation will become.

As our conversation improves, we are likely to view machine intelligence as more than just mathematics, namely as a new category of person, whether friend, colleague, or even family or love interest. Humans don't require much evidence of sentience to project anthropomorphic fears and desires.

In order to see, navigate, and understand us, computers are building a ghostly virtual version of us and of the planet, a shimmering doppelgänger woven together from all of our data. The machines' mental model of the world and ours are beginning to intersect.

I think we are turning the world into a computer, and I will explore how this omnipotent machine 'mirrorworld'[1] is already heavily influencing our lives, and we are only at day one. I will argue that we need to manage these sophisticated technologies carefully, to maximise the huge opportunities they present, whilst mitigating societal and individual risks.

The first half of this chapter looks at the many ways in which sensors are observing and processing, providing the behavioural vocabulary for machines to learn how to understand us. The second half takes one example of how such data might be used, by creating seemingly sentient avatars for us to chat with.

---

[1] www.wired.com/story/mirrorworld-ar-next-big-tech-platform/

Machine Learning and Data Capture

What we think of today as big data will seem quaint in the years to come. Machine-learning technologies are turning every inch of our lives and ourselves into data points. Things we never used to be able to digitally search, like our faces, our journeys, and our heartbeats, are becoming *infinitely* categorised by machines. A rich, illustrative tapestry is being woven for machines to read, contrasted with previously sporadic, loose threads of anonymous information.

We're entering a new data era, an era of visual conversation. It's visual because the conversation is more than just words, emojis, or images. We are chatting constantly to each other and to the machines, passively sending out and receiving signals whether we know it or not. Our data constitute the shared language. Our behaviours are persistently observed by cameras and other sensors, producing massive amounts of new data types that enable the machines to understand us, the world, and our behaviours.

Everything and everyone we interact with, whether cities, buildings, friends, or devices, will have a so-called digital twin, a ghostly shadow self, made up of their data exhaust. These twins, our digital shadows, already influence our lives and will increasingly guide our behaviour. These data will cause us to look at objects and machines differently.

Every cup of coffee, pair of trainers, car engine, or city, will have a halo of data-powered algorithms that will talk to us. What will they say? As they become more alive with connectivity, they will start selling themselves to us,[2] advising us on train times, telling us which direction to walk, or who to avoid. They will tell us when we are displaying early symptoms of a disease or nudge us into healthier and safer behaviours. Menacingly, the nudging could quickly slip into control. As Cory Doctorow suggests,[3] if the manufacturer or tech company can reach down into our home or our body via our devices, perhaps we are our devices' tenants rather than their owners. He warns of a new feudal system.

This watching, listening, and categorising is known as the algorithmic gaze, creating a new reality — a virtual shadow world as seen by

[2] www.diamandis.com/blog/the-spatial-web-part-2
[3] https://boingboing.net/2012/10/22/kindle-user-claims-amazon-dele.html

computers. We are unwittingly training these algorithms; we are the enablers of the addiction. Every time we move, breathe, talk, and walk we feed the hungry beast, adding another brick to the panopticon.

++

Computing is flattening. Having expanded out of large facilities, computing is now seeping into the environment. Tiny sensors loaded with transistors are being deployed everywhere, from outer space to our inner ear and every level in between, turning the world into a computer. I am going to walk through this multi-dimensional sensor diaspora, starting with sensors in outer space and steadily heading down towards solid ground.

If dispersing computing from centralised mega-machines is known as moving to 'the edge', then space is the ultimate edge. Constellations of tiny satellites are being launched into orbit, made up of hundreds of circling mini-computers. These are not just able to send raw data back to Earth, but stream the results, new juicy insights about the planet we were previously unable to deduce.

Over the past few years, this smart swarm of satellites has transformed our understanding of human and natural processes on Earth. We are en route to creating a kind of global CCTV, a searchable resource to answer questions we could previously never hope to answer. Where are all the planes in China? How many trees were felled in Indonesia last week? What are global crude oil levels versus official numbers? Did the election turnout versus election results tally up?

Able to supersede national or corporate estimations, or fictions, these new eyes in the sky have democratised the tools previously available only to the military–industrial complex. Small, privately owned companies like Planet, Digital Globe, or Orbital Insights are giving their clients what feels like 'a direct line to the CIA'.[4]

Observational speed, resolution, and geographical spread increase with each week that passes. More places can be watched, at exquisite resolution, more often. As these resources are being finessed, so increases the

---

[4] https://twitter.com/orbital_insight/status/951114868240785408

range of questions we can now ask. The health of individual fields and likely crop harvest can be extrapolated at scale to forecast potential civil unrest. Planet recently was able to pinpoint unusually high activity at a North Korean chemicals plant at 72 cm resolution.[5] Both of these examples are previously challenging security indicators that have been transformed by remote sensing.

New economic indicators can also be defined from space. Orbital Insights was able to predict the shutting of many JCPenney stores from space, by remotely observing car park activity across thousands of stores.[6] A daily national consumer confidence index could be generated by zooming out and aggregating hundreds of thousands of car parks.

Whether volcano deformation, or pilgrim numbers at Mecca, values for almost any social, political, economic, animal, or vegetable indicator can be obtained, or at least augmented, by information from space.

The question then becomes, what do you want to know? About a company's customers, visitors, colleagues, tenants, processes, logistics? About a government's citizens, assets, landmarks, risks, fleets?

These space-based sensors are radically redefining the nature of risk, and of transparency. If real-time monitoring is becoming possible, the concept of proprietary data is also challenged – is it possible to keep any corporate secrets if satellite imagery is in the public domain? I believe we are entering an era of radical transparency, where companies cannot plead ignorance of illegal behaviour along their supply chains, or in their construction or mining processes. Companies and individuals should assume that they are being observed as much as doing the observing.

++

Drone-collected data provide the next airborne layer of detail; a new tier of sensing at a finer pixel resolution that can augment, or replace, satellite data. Like satellites, individual drones are relatively unsophisticated pieces of hardware, but when swarming they act as a distributed brain. Whether swooping over crowds at Taylor Swift concerts to spot her

[5] www.planet.com/markets/defense-and-intelligence/
[6] https://theoutline.com/post/1169/jc-penney-satellite-imaging?zd=1&zi=ntbereqq

157

stalkers,[7] collecting whale snot over the oceans to examine viruses,[8] or finding meteorite fragments,[9] drones are a new and flexible device for intelligence gathering.

Machine-learning technologies give drones the ability to see and think, and they will soon undoubtedly be able to execute actions as well, like call the police. Louisville, Kentucky, is trialling land/airborne surveillance with lamp-post-mounted microphones, able to detect the make and model of any gun used in a shooting, the number of shooters, and the location.[10] Complementing these static sensors with drones can rapidly add nuance from the scene, and should significantly add to safety. Such concerted hardware collaboration will profoundly add to the depth of data collected, justly worrying civil rights groups like the ACLU. Such surveillance is vulnerable to mission creep.

The visual conversation between us and the drones will be interesting to watch. The Chinese 'Fiercely Wave for Help Identifier'[11] programme suggests that if we want to summon help, we may just be able to wave our hands so that drones, cameras, or other devices can alert the authorities. If the world is becoming a computer, our verbal or physical actions might allow us to command the real world as if it were a machine.

++

Closer to the ground, cars and other robots will provide the next layer of data. As wheeled sensor platforms, cars will contribute more and different data to the global computer. As they transport people and goods, they will observe every street of every city where they are deployed. They will talk about what they see, and the cities will talk back to them. Kevin Kelly describes the way in which robot cars visualise cities as the 'mirror version' of that street[12] or house, its computational reality. Cars will

---

[7] www.theguardian.com/music/2018/dec/13/taylor-swift-facial-recognition-technology-surveillance

[8] www.technologyreview.com/2018/06/06/142452/drones-are-flying-over-whales-and-collecting-their-snot/

[9] https://frontierdevelopmentlab.org/fdl-2016

[10] www.shotspotter.com

[11] https://docs.google.com/document/d/13QqWIFAhNsMnV9uEAc1jK6IBjMzy3c_D_75w7wTcnmc/edit

[12] www.wired.com/story/mirrorworld-ar-next-big-tech-platform/

necessarily generate massive video archives in order to see the world, as they navigate their way around town.

Owners of these vehicle systems will own these archives. Shoshana Zuboff, author of *Surveillance Capitalism*, believes that Ford could switch on a multibillion-dollar new business unit, simply by adding together and selling its car financing and driving data, generated by their massive global fleet.[13]

Data might also give us X-ray vision. Whilst inside a vehicle, the car could collate information collected by residents, office workers, shop assistants, and passers-by and slice open buildings to show passengers what's inside.

Other robots, beyond cars, will also be in pervasive operation, delivering goods, helping the elderly, or serving coffee. Like cars, they will have to be major data gatherers in order to navigate space and interact with humans. For example, the Roomba vacuum cleaner robot takes detailed maps of customers' homes in order to clean autonomously. Roomba's CEO has discussed selling this proprietary data.[14]

Whether cars or vacuum cleaners, robots will be amongst us, acting as pervasive data collectors and creators.

++

There will be many uses for remote sensing by satellite or drone or robot, as machine-learning technologies get better at recognition and processing. Once faces, for example, become searchable, they can be used as ID. I think this is a profound step change from today's still patchy data trails. Once coordinated, your face could become a proxy for your passport, your wallet, or your password.

With our face as the key, we might log into places, shops, restaurants, just like we log into websites.[15] This could transcend one brand or one retailer, and unlock workplaces, streets, even whole cities. Logging or opting out would become an unrealistic option, if accessing services or places relies on assessing your face.

---

[13] www.ft.com/content/7fafec06-1ea2-11e9-b126-46fc3ad87c65
[14] www.theverge.com/2018/4/19/17256074/roomba-irobot-ceo-colin-angle
[15] https://a16z.com/2018/05/14/online-to-offline-china

If this sounds far-fetched, facial recognition and categorisation is a very fertile area of research and investment in tech. In fact, the best-funded AI startup in the world, Sensetime, is a facial data company. To a computer, faces are just a collection of pixels, and those pixels are being crunched in many markets around the world already.

China in particular has a thriving facial data consumer landscape. KFC gives customers menu suggestions that are based on recognition of their faces. Hangzhou no. 11 High School takes attendance, and records library and canteen usage, using kids' faces.[16] Ping An insurance company uses facial emotion software to identify applicants who are lying or require further scrutiny.

Eventually, finding someone might be as easy as googling their name today. I recently gave a talk at Adobe, which announced success in finding missing persons using several data-led technologies. Dementia patients in China are monitored and tracked using facial data to make sure they don't get fatally lost.

One of the most dynamic uses of facial recognition technology today is payments. Amazon Go has proven the viability of sensors and machine vision replacing cashiers and checkout queues, calling their ID-led software 'just walk out shopping'. Further validation comes again from China, with Ant Financial's similar 'smile to pay' technology, and even a Ford/Alibaba car-vending machine that uses customers' faces to purchase a new vehicle.

Persistent monitoring by embedded sensors will enable seamless journeys, purchases, or workdays, with our faces providing the consistency across products and places. Although this would add to customer convenience, help law enforcement solve crimes, or limit fraud, we should exercise caution with such extreme and intimate surveillance.

The vulnerability of technologies like these to mission creep is indisputable. Dystopian scenarios are easy to imagine, some cautioning that we already stand in a 'perpetual line-up',[17] assessed for our propensity for crime moment to moment. Novels and movies, like *Minority*

---

[16] https://uk.reuters.com/article/uk-china-surveillance-education/sleepy-pupils-in-the-picture-at-high-tech-chinese-school-idUKKCN1II128

[17] www.perpetuallineup.org

Report and others, riff on the idea that you could be arrested before committing the crime.

Police forces across the world are trialling facial technologies. Zhengzhou police have deployed glasses that scan for wanted felons at stations or stadia.[18] Police cars in other areas are purported to have roof-mounted cameras that can spot wanted felons even at 120 mph.

Shenzhen uses cameras and billboards to humiliate and prosecute jaywalkers. In an amusing incident showing the limitations of this algorithmic law, Ningbo police recently mis-labelled the head of China's largest air-conditioning manufacturer as a persistent jaywalker, as her face adorned an advert on the side of a bus.[19]

Embedding technologies like these is transforming previously dumb physical infrastructure into a computer, and physical infrastructure is even being replaced by sensors in some instances. Anduril Industries has proposed to outdo Trump's border wall with a virtual wall of sensors, which can identify people, animals, vehicles, and drones, over vast areas to support border control agents,[20] without any concrete.

George Orwell imagined the horrors of persistent surveillance in his novel *1984*, pointing specifically to facial recognition and calling it 'facecrime'. We must be extremely careful not to compromise hard-won rights in the name of convenience and speed.

++

So far I have dwelled on the shrinking concentric circles of observation outside of our homes. The concept of turning the world into a computer extends into the front door, as we interact with sensors inside buildings, and ultimately inside our bodies.

Amazon's Echo was the first significant device to socialise us into speaking to sensors in our homes. What was deemed an intrusion a few years ago is now used by millions of consumers. Although Amazon

---

[18] www.theverge.com/2018/2/8/16990030/china-facial-recognition-sunglasses-surveillance

[19] www.theverge.com/2018/11/22/18107885/china-facial-recognition-mistaken-jaywalker

[20] www.anduril.com

doesn't release sales numbers, they have said Alexa set 100 million timers around the world during the Christmas 2018 period.[21]

Beyond just Q&A, these sensors can also execute actions. Alexa can listen for glass breaking and call the police autonomously.[22] A Cambridge company enables your device to constantly listen for health symptoms and diagnose you.[23] Our homes and buildings listen for instructions, verbal or otherwise, and answer or act accordingly.

Our homes will not just listen, but also watch. Google's Clips is a series of tiny cameras installed throughout a home, trained to autonomously take photos of specific family members or friends. This avoids missing spontaneous moments that are lost in the time it takes to open your phone. Trying to buy Clips was difficult – it was sold-out for months. This points to a future neatly articulated by Siddhartha Mukherjee, where we carry 'some kind of GoPro [video camera], in which data becomes so cheap that you can [monitor] . . . what you do, what you eat, whether you run, how much you run, the number of Fitbit steps, etc. Imagine the density of individuated information that comes from all this.'[24] Dave Eggers explores this eerily dystopian future in his novel *The Circle*.

While these sensors are currently mobile, soon they will be installed in the fabric of buildings themselves. In offices, meetings will be transcribed and summarised and notes circulated to participants by the room itself (companies like Otter.ai do this already via phone apps).

I think we will soon live in the 'Google House' or 'Amazon House'; residential products or systems that we will buy into. Our homes will come pre-loaded. Embedded sensors in walls could perceive someone moving around a house, identify who they are, even read how they are feeling. Professor Fadel Adib's lab at MIT develops these inbuilt sensors, detecting movement and emotion through walls using heartrate and heat patterns.[25]

[21] https://press.aboutamazon.com/news-releases/news-release-details/amazon-customers-made-holiday-season-record-breaking-more-items
[22] https://venturebeat.com/2018/09/20/alexa-guard-alerts-you-when-it-hears-glass-breaking-or-the-smoke-alarm
[23] www.audioanalytic.com
[24] www.nytimes.com/interactive/2018/11/16/magazine/tech-design-medicine-phenome.html
[25] www.mit.edu/~fadel

Our homes might adjust ambient conditions according to who is present, and how they are feeling. Emotion AI is a new field of sensing that will teach algorithms to understand our emotional state and act. New emotion chips could be installed which specialise in reading us and our mood, to allow machines to respond at the same emotional level so as not to anger or upset us.

At the next level down, the falling price of medical testing equipment heralds the ubiquity of sensors that will collate and analyse our biological information, and that of other animals or bugs in our houses. For years we have had sci-fi promises that our bathrooms will read our bowel movements or genetic material from our toothbrushes and recommend or diagnose accordingly. With cheap sequencing technologies like those from Oxford Nanopore,[26] that wish is closer to being granted.

Eventually the so-called last frontier of privacy[27] will be breached, where sensors tap into our brains using neural interfaces. Neuroprostheses will beam our thoughts directly to the machines, to save us the overhead of speaking. Although this is a long, long way off, many big and small tech labs and entrepreneurs from Elon Musk to Facebook have teams working on these devices, but the possibilities and consequences are firmly in the realm of fiction.

++

Despite this multi-dimensional coverage, there will always be data that we can't get hold of. Such data might be confidential, incomplete, or illegal. Developments in generative adversarial networks mean that we can now plug gaps where we don't have complete information. This is a whole new category of data, where libraries of synthetic medical records, molecule structures, or faces are created. Data that can be difficult to access are replicated so accurately that they are as useful as the real thing.

This is tremendously useful for academic research, or anonymising facial datasets. However, if fake news is today's scourge, fake faces, fake

---

[26] https://nanoporetech.com
[27] Bryan Johnson in www.wsj.com/articles/brain-computer-interfaces-the-last-frontier-of-human-privacy-1524580522

people, fake records are upon us. To counter this, projects like Faceforensics are emerging to offer validation of fake versus real.[28]

++

Data are getting deep, and we are training these algorithms. As we go about our daily lives, inside our homes and out, we are contributing pixels to the planetary digital twin. What will we use this for?

I think it will be a tremendous tool to answer questions, reveal radical transparency and accountability, and even predict the future. I believe we are unknowingly building a time machine, a tool with the ability to look back and forwards in time. A data-driven oracle.

As we turn our buildings, airports, homes, offices, and bodies into computers, we can search them for answers, much like we search the Internet. Making sense of all the different types of data is still aspirational. However, with every day that passes, more data and more algorithms will improve that feedback process, and we will learn to ask deeper and better questions.

Imagine a YouTube video, where we can scrub backwards to answer questions about the past. Why did that disease outbreak kill so many people? Why did that area of town get gentrified first? Why did that lake dry up? A palimpsest of data sources will give nuanced results. This could be explored at a city, country, or even continental scale, creating virtual interactive archives of major places of interest.[29]

Continuing the metaphor of a YouTube video, we will also be able to nowcast, from anywhere on Earth. We can ask live questions about activity anywhere. What are people wearing in New York today? How many people are massing on the border in Venezuela? Nowcasting brings planetary stewardship opportunities, holding environmental wrongdoers to account,[30] leaving polluters nowhere to hide, and exposing areas under immediate threat.

YouTube allows you to scrub forward in a movie, to skip to the end. Extending the metaphor, can we then use such a data-driven time

[28] https://arxiv.org/abs/1803.09179
[29] www.timemachine.eu
[30] James Parr, Frontier Development Lab, private conversation.

machine of the planet to extrapolate trends forwards? What might we ask? Movies like *Minority Report* have imagined using such a simulation for predictive policing[31] – where might the next riot break out? Such simulations might enable us to ask what happens if the Internet collapses.[32] What would be the second-, third-, fourth-order consequences? Computer games are excellent proxies for what such simulations might look like in the years to come.

Siddhartha Mukherjee thinks of a 'clinical trial society'.[33] This means the ability, he suggests, to subject aspects of our behaviour to constant trials, using our behavioural and biological data streams. That could lead to many scientific breakthroughs.

The time machine could be made personal: Catherine Mohr asks whether we each could have a 'doomsday clock', an algorithm that shows the immediate effects of eating a certain food or neglecting an activity, on our life expectancy.[34]

The optimists amongst us see these technologies extending our lives, posing questions like is privacy an early death sentence? Unless you share data you are limiting a machine's ability to diagnose you early and often. Mohr echoes this, comparing data to electricity[35]. You don't think about electricity being monitored until something goes wrong; then you are glad it is a metered commodity.

++

As today's newspapers show, how computers view us and our behaviours is not an uncontested field for the techno-optimists. There are many limitations to our growing digital shadows, both societal limits and technical ones.

I will start with a few societal limitations, acknowledging that there are many others I don't have the wordcount to include.

[31] https://journals.sagepub.com/doi/10.1177/2053951718820549
[32] https://improbable.io/games/blog/what-we-found-when-we-simulated-the-backbone-of-the-entire-internet-on-spatialos
[33] www.nytimes.com/interactive/2018/11/16/magazine/tech-design-medicine-phenome.html
[34] www.nytimes.com/interactive/2018/11/16/magazine/tech-design-medicine-phenome.html
[35] www.nytimes.com/interactive/2018/11/16/magazine/tech-design-medicine-phenome.html

Intimate data are very powerful. As our photos, searches, footsteps, and lightbulbs become carriers of information, and our faces evolve into our wallets, our passwords, our passports, even our alibis, the power in the data becomes very apparent. The emerging field of 'data ethics' is an early attempt to help regulators, citizens, and companies educate and police themselves.

Many are sounding a warning against abstraction – after all, 'big data is people.'[36] When data can affect our lives in meaningful ways (whether we are approved for a loan, get into a school, secure bail), then we have to be hyper-vigilant about the responsibility of those who collect these data and how they use – or exploit – them.

This becomes even more serious when you don't think the machine truth and your own truth match up. As artist Trevor Paglen's work demonstrates,[37] algorithms can dangerously miss the target. Whilst an algorithm mistaking the subject of a Manet painting for a burrito might be amusing, machines mis-categorising data could have detrimental consequences at an individual level as well as a systemic one.

Beyond categorising us, civil liberties groups are justly concerned about our autonomy in a data-driven world. Once embedded in our systems, machines are not just viewing and processing us, they are nudging us into different behaviours. In some economies, that is a government-driven approach, as in China, where the social credit score encourages citizens away from behaviours that the government doesn't approve of.

In more open societies, consumers can be unwittingly nudged into buying objects or services. Drivers already exhibit different behaviours if their insurance company instals a 'black box' in their car to collect data on their driving nuances.

Pokemon Go is a useful first test-case of this future. The real world, in this game, is augmented with virtual data. Virtual creatures are superimposed onto the real streets that you view through your phone, to incentivise play. This sort of game heralds the creation of what some are

---

[36] www.oreilly.com/radar/why-we-need-to-think-differently-about-ai/
[37] https://qz.com/1103545/macarthur-genius-trevor-paglen-reveals-what-ai-sees-in-the-human-world

calling a 'virtual behavioural futures market',[38] paying not for our eyeballs but for our physical presence. Adverts floating above shops ('Sophie – remember you need to buy toothpaste, buy it here at 2 for 1'), restaurants ('Sophie, have lunch here, your sister ate here in August and left a virtual note that you would like it'), or museums ('the FT gave this exhibit 5 stars') would be driven by our data. Such virtual real estate would be bought in a traded marketplace.

This pay-to-play model, where companies pay to be highlighted in your camera's viewfinder, is beautifully presented in Keichii Matsuda's dystopic short film *Hyper-Reality*.[39] Matsuda's imaginative film features a world annotated by information that nudges us towards places, products (like Coca-Cola), people (maybe potential friends), or actions (like weight loss).

Huw Price and Karina Vold at Cambridge talk about 'sanctimonious silicon curtailing our options', arguing that these algorithms would turn places, or cities, into a 'well-curated moral zoo' where we 'don't notice the fences'.[40] When algorithms are making our decisions for us, smart cities limit individual agency and democracy.

The ownership of the data is another sensitive area currently under debate. As citizens wake up to the level of surveillance from objects, places, and searches, the complex concept of data ownership surfaces. Can a company own what I see through my eyes? Where is the IP in my data? Are the heartbeats from my pacemaker data mine? Who will own the data inside autonomous vehicles?

Tim Berners-Lee offers a solution to some of these concerns. He believes in the re-decentralisation of the web and has developed a software platform to work towards his original vision for his invention. Berners-Lee would like to see us communicating without Big Tech middlemen. His platform Solid[41] is a way to create a personal online data store, which gives the user control over who accesses their data. The blockchain has also been proposed as a method to provide a secure layer between the digital and physical worlds.

---

[38] http://shoshanazuboff.com/
[39] http://hyper-reality.co/
[40] www.cam.ac.uk/research/discussion/living-with-artificial-intelligence-how-do-we-get-it-right
[41] https://solidproject.org/

Regulators are accused of trailing behind, of a yawning 'law lag' as technologies seemingly accelerate away from policymakers' capabilities. Europe's GDPR, and recent Californian legislation, are steps to protect personal privacy in law. It is a highly politicised arena. *Foreign Policy* magazine accuses Facebook of exaggerating Chinese citizens' comfort with surveillance, to convince US policymakers to go easy on Facebook's data aspirations. By spooking policymakers into fear about Chinese tech domination, Facebook is accused of suggesting that privacy rules limit innovation.[42]

These, and many other, societal concerns and opportunities require more research, new roles in data ethics and beyond, and a greater investment in skills.

++

Technically, there are also limitations to these data ambitions.

I recently co-founded a company at Oxford University,[43] to tackle a blockage in the data supply chain. Without clean data, algorithms can't find patterns or execute tasks. Companies and governments sit on years' worth of data stockpiles which have not been labelled or organised. Data quality is a big bottleneck in the promise that such data provides. AI technologies radically improve with more and better data.

New chips are also required as we near the end of Moore's Law. Soon, classical physics will limit the price–performance curve that has sustained the past 40+ years of technological innovation. New chip architectures are needed for the complexity of new algorithms, and to support the move from centralised computing to the ubiquity of tiny computers I described earlier in this chapter.

Moreover, we will probably also need an entirely new computer architecture. Machine learning and other technologies will require new types of processing, smarter than brute force. Quantum computing is one example of a novel architecture; with its speedy optimisation and reduced computational complexity, it is perhaps more suited to algorithms than today's classical machines.

---

[42] https://foreignpolicy.com/2018/08/14/the-data-arms-race-is-no-excuse-for-abandoning-privacy/
[43] www.1715labs.com

Data storage is also a question – as we collect ever more data, where will we put all these data? Some of the intelligence will remain on the edge, on the device where it's collected. This will also be more secure. However, we will still need centralised facilities, which tend to be heat- and energy- intensive. Microsoft has experimented with undersea data storage[44] to provide a cool, renewably powered environment; others even propose space as the ultimate cold storage site.

New ways to shrink data storage are being explored, and DNA is one frontrunner. We know that under the right conditions DNA can store information for millennia. Microsoft has already trialled using genetic material in their data centres as a miniaturised storage mechanism.

++

We may think we know big data, but I think we have only just begun. What we think of as big data today will seem like tiny data in a few years' time.

Data feed how the machines view, understand, categorise, and even nudge, us and our behaviours. We are building a time machine where we can use these data to understand the world – and ourselves – better, but this ability could also be used irresponsibly, putting civil rights at risk.

I think we should be actively working on managing data-collection technologies now, to leverage the enormous opportunities such radical transparency brings (from early diagnosis, to illegality in supply chains), but mitigate the major downsides associated with stepping into a '1984-style' surveillance universe. We have to work hard to ensure we are not 'Huxleyed into the full Orwell'.[45]

Sentient Avatars

The machine-learning technologies that process all of these data arise from the artificial intelligence family.

---

[44] www.theverge.com/2018/6/6/17433206/microsoft-underwater-data-center-project-natick
[45] https://motherboard.vice.com/en_us/article/8qxdy4/huxleyed-into-the-full-cory-orwell-cory-doctorow

The second part of this chapter will focus on how these data might give life to the machines. I will explore what it might feel – and look – like to converse, confess, and transact with algorithms, or even be made immortal by algorithms.

It is fashionable to be a doom-monger, or a wild optimist, about artificial intelligence. Wherever you are on the spectrum, there are big questions to consider. Some of the biggest debates on this topic are existential, as synthetic intelligence is championed by some as a bio-logical upgrade for our species. Debates rage on whether the next rung on the human evolutionary ladder will be a silicon one. This has led to wild hype on the investor circuit, and a wild media and book-writing circus forecasting the end of humans.

In my view, we will be surrounded by myriad discrete intelligences. They will be robots that are very good at a small number of specific tasks but show little promise in gaining any universal skills. As these servants perform tasks for us, they will augment us both at work and at play.

Cumulatively, however, their collective actions might feel like superintelligence long before they actually are. When Alexa turns on your lights, she doesn't know what electricity is, what a light is, who 'she' is; she is simply performing a series of discrete instructions. That is not yet intelligence, yet could feel like it to those who aren't aware of its technical shortcomings.

Most technical experts don't worry about an imminent synthetic intelligence explosion. The general consensus is that so-called artificial general intelligence (or AGI) is within a 25–40-year horizon, rather than an immediate threat.

However, I think that a software universe that can approximate the feeling of AGI is almost as important, because the human pushback and fear levels will be identical to what they would be if it were the real thing. The perception of all-knowing AI might be felt long before it is an engineering reality. As Professor Hans Rosling talked about in *Factfulness*,[46] perception can be as important as reality.

Misinterpreting the output of an algorithm, or its statistical range, is likely to be a common occurrence. Although many technical experts scoff

[46] www.gapminder.org/factfulness-book/

that AI today is not really intelligent but merely 'advanced statistics', most people don't understand the basics of the maths that underpins predictions made by machines.

++

As our thoughts are processed as much by machines as by our brains, and as we begin to make major life decisions using algorithms, we should assess how comfortable we are with these innovations. I agree with Ezra Pound's view that 'artists are the antennae'. I find imaginative future visions can be a useful guide on how to act responsibly as companies, and as individuals.

Being observed by machines in our homes is a profound imposition. Lauren McCarthy's work *Get Lauren*[47] is a provocative performance piece where she installs herself on a screen in your home. She requires us to ask whether we are happier with a real person (Lauren herself) watching us in our house constantly and answering our questions, versus identical data being held in a data centre and owned by a company.

How we will feel as we have human-like conversations with robots needs to be examined. Rekimoto Labs trialled a 'chameleon mask',[48] an iPad-style tablet placed over a surrogate person's face livestreaming the face of a second – remote – person, who navigates them around a workplace or home. They found that people treated the surrogate as the person whose face was superimposed on theirs. Might robots then simply require a realistic face imposed on them for us to treat them as a human?

It appears that some small interventions can help humans feel more relaxed around robots. 'Boxy'[49] is a robot that uses its unthreatening cardboard exterior and smiling face to instil confidence in humans. Asking for directions and deliberately crashing into walls are pre-programmed 'vulnerabilities' that allowed human passersby to assume Boxy is helpless and unthreatening.

These – and many other performance pieces and experiments – creatively demonstrate the realities of the fledgling relationship between

[47] https://get-lauren.com/
[48] https://lab.rekimoto.org/projects/chameleonmask/
[49] http://areben.com/news/

man and machine. Machine intelligence is blurring the boundaries between real and not real, human and non-human. The new computational reality that we are moving into requires a new ontology to describe how we see the machines, and how they see us. Iyad Rahwan and Manuel Cebrian at MIT[50] propose a new academic field akin to anthropology or sociology, called 'machine behaviour'.

It is in our interests to understand machine behaviour, as we need to leverage the differences between our intelligence and other types of intelligence. We have strengths against the machines, and they have strengths against us. In order to solve some of the major problems we face – whether climate change, pandemic management, or financial crises – we will need to work together to optimise both types of intelligence. It should be a race with, not against, the machines.

How might we work with them? I want to illustrate this with one example of avatars, digital entities that we might interact with, exactly as we currently interact with humans.

++

Imagine a digital double,[51] a virtual model of you, based on algorithmic models that companies use today, loaded with our preferences and activities. That model gets increasingly more detailed with time. As it becomes more robust and representative, we might send it out into the world. Instead of us browsing Amazon, our representative might do that for us. It could replace us in the digital economy, like an ambassador in foreign policy negotiations.

The key part is that this would not be supervised by us. Like ambassadors, these bots will be autonomous, making their own decisions to our perceived benefit. We will allow them to make decisions and purchases that we don't have the time or knowledge to do ourselves. Ephemeral avatar versions of us will float around the Internet, bringing us efficiencies and expertise that were previously impossible.

---

[50] http://nautil.us/issue/58/self/machine-behavior-needs-to-be-an-academic-discipline
[51] www.scientificamerican.com/article/artificial-intelligence-will-serve-humans-not-enslave-them/

They will also bring us the Internet. Tech CEOs like Satya Nadella at Microsoft expect that voice agents will replace the web browser: they will be the 'prisms through which we interact with the online world'.[52] Our interactions with them will make the Internet more alive.

Early signals of this sort of shift are under way. Google's 'Talk to Books'[53] platform enables us to ask 100,000 books any question we like. Examples might be 'Do AIs have consciousness?' or 'Where shall I hike in the Canadian Rockies?' Answers are selected from thousands of human-written books, and quotes from authors on the topic. It is a great idea-generator. It is an early instance of being able to tap into the collective intelligence of our species, like having a conversation with Wikipedia rather than reading it.

You could imagine doing that with your personal archive – asking your algorithm 'What should I get my father for his birthday?' or 'Why am I depressed today?' Your own life, and your own cumulative intelligence, might store in silicon the answers to those questions, which aren't easily recalled using our wet, organic brains.

Another early signal of a change in our relationship with data is being socialised by digital online personalities like Lil Miquela[54]. Miquela's graphics are so lifelike it is not always obvious that she is essentially a cartoon. On Instagram she wears high fashion, is sponsored by haircare brands, has an agent, and has done magazine shoots. Her 'skin is flawless'[55] according to one of her 1.5 million fans. A modelling agency called the Diigitals, with a roster of digital models available to advertise clothes or skincare, has been launched. A particularly 'meta' moment was reading about a 'real' model's experience standing in for the digital model's body[56]. Such 'authentic fakes'[57] are seamlessly merging with people's *real* fake lives on Instagram.

---

[52] www.slate.com/articles/technology/cover_story/2016/04/alexa_cortana_and_
siri_aren_t_novelties_anymore_they_re_our_terrifyingly.html?via=gdpr-consent
[53] https://books.google.com/talktobooks
[54] www.instagram.com/lilmiquela/?hl=en
[55] www.instagram.com/p/Bst1fe4Hjy4/
[56] www.thediigitals.com/misty
[57] www.standard.co.uk/lifestyle/esmagazine/lil-miquela-ai-influencer-instagram-
a4084566.html

What are these early signals leading to? Some propose that an 'avatar layer'[58] will emerge, where every company or service has an avatar with deep knowledge of its business, that we will speak to (or indeed our avatars will). Banks, cars, shops, and offices will have infinitely scalable avatars to chat with. In a virtuous circle, the data unlocked by this chatting will be incredibly deep, in turn feeding the avatars' data vault.

In the workplace we will use these avatars to begin, and advance, our careers. Early in our lives our avatars might comment on our propensity for success in certain careers and not others. A quick scan of a lifetime of emails and texts might demonstrate a skill for negotiation but not subtlety, for example. Once career tracks have been identified, we might ask our avatar to attend interviews on our behalf for hundreds of jobs, coming back with a shortlist of jobs we are suitable for, and that are suitable for us.

Once in a role, we could call on the expertise of others, using their avatars. MIT's Media Lab has a project called Borrowable Identities,[59] where we might summon a subset of the intelligence or opinions of Ruth Bader Ginsburg, for example, if we were struggling as a trainee lawyer. Some suggest that heads of state or CEOs could use such a process to ensure smooth succession planning.

In the workplace, we could collaborate with our colleagues – or with friends or strangers – to complete certain tasks. 'Have your AI cc my AI'[60] might be a regular refrain. Questions about how companies start to deal with autonomous hires and autonomous consultants will start to arise today.

Kevin Kelly believes that we might wake up in 2043 and find that a million people have collaborated successfully to build something,[61] amateurs working together to solve real-world problems collectively. He believes we are at the absolute beginning of all of these avatar-type developments.

---

[58] Armando Kirwin in https://singularityhub.com/2019/01/15/the-rise-of-a-new-generation-of-ai-avatars/
[59] www.media.mit.edu/projects/augmented-eternity/overview/
[60] www.slate.com/articles/technology/cover_story/2016/04/alexa_cortana_and_siri_aren_t_novelties_anymore_they_re_our_terrifyingly.html?via=gdpr-consent
[61] www.wired.com/story/wired25-kevin-kelly-internet-superpowers-great-upwelling/

I think they will also help us with our personal finances. Being able to talk to our finances might be a huge benefit to people who currently rely on middlemen to translate. Can I afford that holiday? What impact would that pay rise have on my mortgage? What would my retirement look like if I saved less in my pension? Virtual reality could help show very detailed simulated scenarios.

Healthwise, avatar interventions could be very beneficial. Pedro Domingos suggests that they might assess options for us upon diagnosis, given they will be likely to know our genetics, our behaviour, our family history, our propensity to take pills, and our budget, and recommend therapies. Domingos goes as far as suggesting that we might put forward these avatars for medical research, if they truly represent us and our most intimate data.[62]

Mental health therapies are being trialled now with avatars. The Woebot,[63] from Andrew Ng's team, has spoken to millions of people, providing therapy particularly to populations who wouldn't ordinarily access it. For some, speaking to an avatar is better than a human, as they feel less judged or stigmatised.

Doubtless, as the movie *Her*[64] so beautifully illustrates, we will fall in love with avatars too. They will finally understand us in ways that real humans aren't always able to. Not much sentience is actually required for us to imagine a satisfying relationship. Enough people propose to Alexa every day that she even has a pre-programmed response.

Some even think that we will send these avatars into deep space, given the impracticalities of sending humans. Perhaps future spaceships will talk to us, sharing what they see from deep in our galaxy.

The community of avatar watchers believes we'll use them to live out hundreds of versions of our lives, so we can experience the best possible human version of our real life. However, if they are going to represent us in such a deep and personal way, we need to ensure that they embody our

[62] www.scientificamerican.com/article/artificial-intelligence-will-serve-humans-not-enslave-them/
[63] https://woebot.io/
[64] https://en.wikipedia.org/wiki/Her_(film)

values.[65] As Soul Machines' 'Baby X'[66] project hints at, we might grow with 'our' AI from birth, so we had better shape it to achieve our goals.

++

Beyond having autonomous algorithmic people in the economy, there will also be autonomous algorithmic companies, companies made solely of software that will buy and sell services, products, knowledge, and skills to other machines, and to us.

We might eventually be able to talk to a company and ask it questions. Can you draw up this contract for me? Do these rules apply in my jurisdiction? Will you put up prices next quarter?

There is great debate about the legal personhood of robots, which is further complicated by autonomous companies. If a robot has control of a limited company, should it get the right to own property, sue, and hire humans? In the meantime, the Competition and Markets Authority has already investigated[67] instances where companies' algorithms are behaving like cartels, exhibiting anti-competitive behaviour even without the knowledge of company directors. These are all questions that will require increasing examination.

++

I'd like to introduce a few ideas to consider about these future, virtual, actors in our lives.

Firstly, companies that advertise directly to my avatar might gain competitive advantage. If I have outsourced many of my purchasing and decision-making tasks to an algorithmic version of me, showing me advertisements might be superfluous. For certain categories of purchasing, marketing to the company which runs my avatar might be more effective.

Secondly, who owns or maintains our avatars? Given the depth of data required to make these avatars effective, it is important that we understand their data policies. Designer Francis Bitonti described to me the

---

[65] https://medium.com/@johnsmart/your-personal-sim-a07d78ffdd40
[66] www.soulmachines.com
[67] www.gov.uk/government/news/algorithm-research-builds-on-work-in-digital-markets

horror of an *extreme filter bubble,* when these bots don't just represent us to the world, but represent the world to us. Sergey Brin, co-founder of Google, said he hoped Google would be the 'third half of our brain'.[68] Maybe the avatar is this digital third half? Sherry Turkle at MIT cautions that avatar behaviour raises ethical questions. Our 'Darwinian buttons'[69] are pushed when these avatars manipulate our emotions by asking us intimate questions.

Thirdly, as the Google Transparency Report demonstrates, information about its users is handed to the police. Previously, this type of information would have required a search warrant to access – photos, receipts, calendar entries, correspondence. In the second half of 2017, there were 3,773 disclosure requests by government agencies in the UK.[70]

Science-fiction scenarios emerge when you flip this question, and ask who would be responsible if an avatar were to commit crimes? With the knowledge of its owner, and without.

Fourth, some believe that there is no reason why these avatars couldn't, on occasion, be embodied. Imagine Siri or Alexa getting 'a virtual, 3D, photorealistic shell, whether machine, animal, human, or alien'. Kelly calls this the 'badly needed interface where we meet AIs, which otherwise are abstract spirits in the cloud'.[71]

If they were to join our conversations, that would also be a new experience. Might they fact-check our statements? Interrupt us? Remind us to be more open-minded?

The final idea I'd like to posit is that these avatars might offer us a way to live forever: digital immortality. Futurist and Googler Ray Kurzweil, and others, argue that we are information, and therefore creating an avatar version of us will enable us to back ourselves up, create a silicon version of our sentience for eternity. Could software give us digital heirs?

Odd as this might seem, the web is increasingly 'inhabited by the remains of departed users'. Facebook and other social networks are profiling both

---

[68] www.businessinsider.com/sergey-brin-we-want-google-to-be-the-third-half-of-your-brain-2010-9?r=US&IR=T

[69] www.dhi.ac.uk/san/waysofbeing/data/communities-murphy-turkle-2007.pdf

[70] https://transparencyreport.google.com/user-data/overview?hl=en&user_requests_report_period=series:requests,accounts;authority:GB;time:&lu=legal_process_breakdown&legal_process_breakdown=expanded:0

[71] www.wired.com/story/mirrorworld-ar-next-big-tech-platform/

the lives and the afterlives of many of us. There is a thriving digital afterlife industry that will only increase with time. MIT's project Augmented Eternity allows you to 'create a digital persona that can interact with people after you're dead'.[72] By selecting the digital persona of anyone in your address book, you could chat to them as if they were alive.

We may also consume entertainment by deceased performers alongside live stars. Actors and movie studios routinely scan all their leads, the Star Wars franchise in particular. Visual effects specialist Beau Janzen has said 'It's inevitable that casting deceased actors will become more prevalent.'[73] The murdered rapper Tupac famously played at Coachella music festival in 2012 using holographic technology.[74] Brazilian rapper Sabotage's estate went one step further after his death, entering a collaboration with Spotify. They used an algorithm to create IP from his back catalogue and released a new piece of original music.

All of this leads to questions about rules and regulations protecting our digital afterlives. Carl Öhman and Luciano Floridi at the Oxford Internet Institute have created a framework to help us think these questions through, which is based on the ethical standards for handling physical remains. Their proposals limit the ability for companies to posthumously exploit personal data, calling it the 'informational corpse of the deceased'.[75]

I think that the Internet of the future will be partly inhabited by our ghosts. We will become used to seeing and interacting with both live and dead friends, family, and public figures, as well as algorithmic, synthetic people, all side by side.

++

To conclude, the radical transparency brought about by data has enormous potential; making our lives more efficient, more accountable, and healthier, with early and regular diagnoses. However, many of these data collection and analysis technologies are especially vulnerable to mission

[72] www.media.mit.edu/projects/augmented-eternity/overview/
[73] www.technologyreview.com/2018/10/16/139747/actors-are-digitally-preserving-themselves-to-continue-their-careers-beyond-the-grave/
[74] www.youtube.com/watch?v=-FGjbCgu-PM
[75] https://papers.ssrn.com/sol3/papers.cfm?abstract_id=3172038

creep, and we should bake in safeguards for our hard-won rights and protections, especially for the disenfranchised.

Data-driven avatars will be the first of many robots in our lives. Whether they are our sanctimonious overlords or hyper-efficient assistants is up to us. Philosophers have considered human purpose, mortality, and identity for centuries, and the advent of conversational robots is the next chapter of those discussions. We should listen to ethicists, sociologists, zoologists, and a host of other thinkers, to help us define our place in a machine-enabled future.

A different interpretation of the word 'vision' is implicit in this essay and to my job as a futurist, and in the creativity of the sci-fi writers, makers, hackers, and gamers I am lucky to spend time with. I believe that we should spend more time imagining the future.

We don't, as the saying goes, suffer from a lack of technology, but instead from a lack of imagination regarding what to do with it. Corporate and government decision-making is often short-termist and rushed, and we risk sleepwalking into a dystopian future. Vision and imagination might help to guide a more responsible use of technologies, pointing these new artificial superpowers at planetary challenges that so far we have been unable to solve alone.

7 Computer Vision

ANDREW BLAKE

Computer vision is a science that has lately come of age. Professor Li Fei-Fei, as Chief AI[1] Scientist at Google, recently went so far as to characterise computer vision as 'AI's killer app' [1]. There might be some overstatement here – clearly speech understanding, language translation, and text search are AI killer apps too. Indeed, Professor Li is known for her optimism about computer vision: in her 2015 TED talk[2] she implied that AI vision was already on a par with the vision of a three-year-old child: 'To get from age zero to three was hard.' But it is true that computer vision is now working well, solving certain problems that have real commercial and human value. And it was indeed hard to get there, taking 60 years or so of theorising and experimentation in the laboratory, but taking most of that period to close in on working vision.

A rough timeline for computer vision (Figure 7.1) begins in 1956 when AI was born, at the Dartmouth workshop [2]. Turing Award laureates John McCarthy, who went on to pioneer AI at Stanford, Marvin Minsky, who soon went to MIT, and others, ran a summer of AI at Dartmouth College. An early foray specifically into computer vision came in the early 1960s with the work of L. G. Roberts [3] (who went on also to be a leading contributor to the development of the Internet). He took images from a camera and processed them on a computer, for example using differencing operations to enhance the edges of a simple polyhedral block. Soon after this, Papert and others at MIT launched their

---

[1] Artificial intelligence.
[2] www.ted.com/talks/fei_fei_li_how_we_re_teaching_computers_to_understand_pictures/transcript#t-1036935

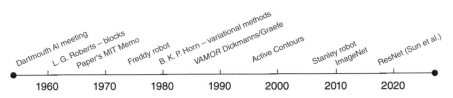

FIGURE 7.1 A timeline for computer vision. See the text for details.

famed summer project [4] to teach computers to see, reckoning that simple objects such as blocks could be dealt with in July, and more complex ones would take until August. This proved to be overly optimistic, and researchers are still working hard on these problems more than 50 years later. Early vision-guided robots, such as the Edinburgh Freddy robot [5] were surprisingly able, if a little slow, by the mid 1970s. The 1980s saw the development of the first convincing mathematical foundation for vision, with the work of Berthold Horn at MIT [6]. Autonomous vehicles, using vision to drive at speed, arrived in the late 1980s with Ernst Dickmanns' and Volker Graefe's Vamor van [7]. They went mainstream, in research at least, when, in 2005, Stanford's Stanley, an instrumented Volkswagen car, drove more than 100 miles across the Mojave Desert to win the DARPA grand challenge[8]. Mathematically based algorithms for capturing shape – so-called active contours – matured in the late 1990s [9, 10]; but, in the new millennium, mathematical models began giving way to big data, notably with the ImageNet challenge [11], which went on to drive substantial advances in vision systems that learn. One final landmark has been the deep neural network (DNN) revolution, which began to transform computer vision in 2012 [12], and is developing apace, such that now deep networks consisting of 100 or more stages are commonly in use [13].

Computer vision has had a remarkable evolution over the course of 60 years or so, and that is what this chapter aims to illustrate, in three parts. First, there is a celebration of four different computer vision systems that really work. Secondly, three fundamental principles are discussed, to give at least a sense of how vision systems work. Thirdly, in a kind of confessional, three aspects of computer vision for which there are still mysteries to solve over the next decade are discussed.

Four Types of Machine That See

Computer vision is now a mature science, and examples abound of machines that have effective vision, and are ready for commercial use.

Human Body Motion Capture

Camera systems pointing at a moving human body, and feeding video to a computer, can be used to compute how that body is moving in some detail and with good temporal resolution. This has many applications, in movie special effects, for video games, in orthopaedic surgery, in physical security systems, and elsewhere. Earlier systems such as Vicon[3] achieved reliable operation by laborious instrumentation of the body, requiring affixment of markers or the donning of a special suit. This is how the character of Gollum in *Lord of the Rings* was brought to life, the artificially generated creature moving in lockstep with the movements of actor Andy Serkis.

A harder challenge is to do this *without* the need for preparatory treatment of the body, applying computer vision analysis directly to the natural appearance of a moving person, with tolerance for variability of size, shape, and clothing. This was achieved on a commercial scale with the Kinect 3D camera system [14] that used machine learning (see later in this chapter for more details). It was launched in November 2010 and set a new record as the fastest-selling consumer electronics device of all time. Its operation is illustrated in Figure 7.2, where a three-dimensional video is analysed by computer in two stages, eventually producing a stream of coordinate positions of the main joints of the body, over time.

In the decade since Kinect was invented, many improvements have emerged in body motion tracking, and most recently a remarkable milestone was reached in the development of an entirely new kind of unmanned shop. Amazon Go is a shop with no checkouts, where the activity of shoppers, their movements, and their interactions with goods on the shelves is monitored so effectively that a shopper can simply pick items and walk out of the shop, and be charged automatically for their purchases.

---

[3] www.vicon.com

FIGURE 7.2 The Kinect human body motion capture system. The Kinect 3D camera uses active stereoscopic vision to capture a depth-map of the body moving in its field of view. The moving depth-map is then relayed as a video to a computer, which analyses it using ensembles of decision trees to label about 30 different areas of the body, as shown by the harlequin colour scheme. This approximate map of the body and its parts is then fed to a more detailed body that ascertains the positions of joints in each successive video frame.

Faces and Emotions

Vision technology has reached the point that faces can be detected in images, identified, compared, and analysed with considerable accuracy and reliability. In the early 2000s scientists at Mitsubishi's laboratories in Boston developed a programme that learned to recognise faces in video efficiently and effectively [15]. Until then researchers had generally assumed that it would be necessary to 'track' faces through successive frames of a video, using the estimated position in one frame, together with an estimate of the velocity of any movement, to prime the search for the face in the next frame. The Mitsubishi scientists remarkably demonstrated that they could detect faces in each frame of a video with no need for tracking at all.

Once a face has been detected there is much more that can be done. The technology of recognition of individuals has advanced considerably, including recognition from scratch and maintenance of identity over time. Analysis of individual facial features or landmarks has become reliable, thanks in part to active contour algorithms (mentioned in the timeline section earlier in this chapter). This allows emotional responses, such as fear, surprise, and anger, to be detected and also to be quantified [16], as Figure 7.3 illustrates.

Computer Vision in Medicine

The use of vision machines for analysis of medical diagnostic images is an entire field nowadays, with more than one major conference to itself. In

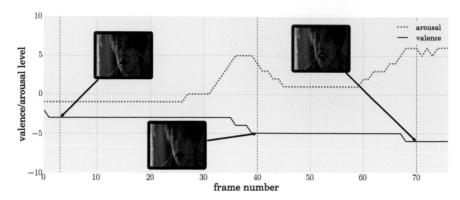

FIGURE 7.3 Visual analysis of emotional response. Automated analysis of facial features in a video allows the expression in each frame to be quantified in terms of degree of arousal, and also 'valence' – a measure of how positive an emotion is perceived to be. Figure reproduced by kind permission of the authors of [16].

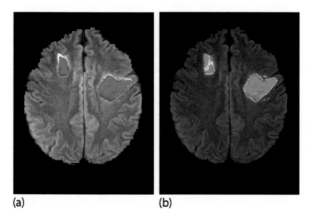

FIGURE 7.4 Identifying a tumour in a brain scan. Reprinted by kind permission of the authors from [17]. Image (a) shows tumours identified by a radiologist, and image (b) shows the corresponding delineation by a computer programme using a 'convolutional neural network'.

particular, vision is having a big impact on radiology. There is now an annual conference, CAR, dedicated to AI in radiology, including CT X-ray, magnetic resonance imaging, and ultrasound measurements. Computer programmes like the one employed in Figure 7.4 are able to locate and delineate semi-automatically clinically significant features such as tumours. This is particularly important when planning surgery on a brain tumour [18], where it is critical to remove just enough material, but no more, in order to minimise impairment of brain function. Progress has been sufficiently impressive to prompt Geoffrey Hinton, one of the originators of deep learning, to say, perhaps somewhat in jest, 'The role of radiologists will evolve from doing perceptual things that could probably be done by a highly trained pigeon to doing far more cognitive things' [19].

Driver Assistance and Autonomous Vehicles

The pioneering achievements of the late 1980s [7] and in the current millennium [8] have raised hopes and expectations of vehicles that will navigate themselves, even in city traffic. Already in the early 2000s automobile companies such as Mercedes were experimenting with visual driver aids such as automatic braking for pedestrians [20] and now a number of manufacturers offer capabilities such as car-following, active cruise control, and visual lane keeping. In the Society of Automobile Engineers (SAE) classification of degrees of autonomy [21], this sort of driver aid ranks as 'level 2'. Levels reach to 5, which is complete autonomy – cars without steering wheels or pedals. It is currently unknown when or whether complete autonomy is attainable, but the prospects for level 4 autonomy – operation without any driver involvement but restricted, for example, to certain geographical areas, seem promising. Numerous companies, including large ones, such as Google and Intel, and startups like FiveAI[4] in the UK, are building the technology for level 4 autonomy. These are complex AI systems with many moving parts, but, of course, computer vision plays a central role, as illustrated in Figure 7.5.

Three Principles for Vision

Building a machine that sees is deceptively difficult – deceptive because we humans do it so effortlessly, using the very considerable, highly parallel, computing power contained in the cranium of each of us. The difficulty is illustrated in Figure 7.6. Computers tend to take data rather

[4] https://five.ai/

FIGURE 7.5 Autonomous driving. The view in the 'brain' of an autonomous vehicle, assessing the topography ahead and allowing for the presence of any obstacles to scope out the safe driveable area. Reprinted by kind permission of the FiveAI company.

FIGURE 7.6 Computer vision is hard. The aim here is for the computer to perceive the outline of the hand in the centre, and the result on the right seems obviously correct and in line with our own perceptions. However, that result has been produced by a carefully designed 'active contour' algorithm with a good deal of prior knowledge about hands [9]. A more literal reading of the data on the part of the computer looks more like the image on the left. The true contour of the hand is confounded by gaps, skin texture, shadow, and background clutter, all of which, on reflection, are indeed present in the picture. How is it that human vision or machine vision can effortlessly ignore the extraneous detail?

literally, and that introduces confusing detail that confounds the clear interpretations that seem obvious to us.

Probabilistic Mechanisms

Human vision quietly deals with the ambiguity in images on a substantial scale. This is well known to visual psychology, for example in the visual illusions [22] in which the interpretation of an image flips between multiple states – for instance convex to concave; vase or faces; young model or old woman. Other illusions depend on camouflage, such as the famous Dalmatian in the trees, where an interpretation emerges only after scrutiny, during which the visual system assimilates a forest of noisy and competing features. It seems that the human visual system treats an image as a complex field of fragments of evidence, against which, rather like a scientist, it generates and considers various hypotheses [23] until it finds one or more that explain the evidence best. Therefore, in computer vision systems it is natural to work not so much with the arithmetic that computers normally perform, but in terms of computation with probabilities – probability, after all, is the natural mathematical calculus of uncertainty.

Probabilistic computation is ubiquitous in computer vision systems. One simple way in which images can be modelled using probabilities is to regard the colour/intensity of each pixel in an image as a random variable. That approach can be used to separate the foreground of an image from its background (Figure 7.7) by acknowledging two probabilistic tendencies in the pixels: spatial coherence and adherence to palette. Spatial coherence refers to the tendency of foreground pixels to form in clumps, and similarly the background. It is represented by a so-called 'Markov random field' [24]. Palettes are built up to represent the distribution of colour and intensity that characterises a particular foreground and background. Combining coherence and palette via a suitable optimisation algorithm is sufficient to automate, at least in part, the separation of foreground from background [25]. Microsoft has used this principle for the 'background removal' tool in all versions of Microsoft Office since 2010. This is just one example of the way in which probabilistic modelling can be used in image processing, and nowadays it is a ubiquitous tool for machine vision.

Learning by Example and Big Data

For a couple of decades, in the 1960s and 1970s, there were many attempts to programme machine vision systems in terms of rules and sequences of operations on images – rather in the manner of step-by-step recipes. Then, in the 1980s, through the work of pioneers like Berthold

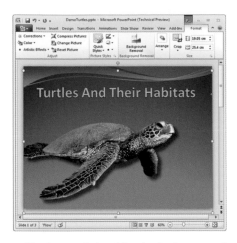

FIGURE 7.7 Separating object and background. An object in the foreground is re-sited to a new setting. Probability distributions are used to model palettes of colours for each of the foreground and background in the image on the left. The distributions can then be used in automated probabilistic reasoning, using also the prior knowledge that lumps of matter tend to be coherent. Reprinted by kind permission of the FiveAI company.

Horn (see the timeline in Figure 7.1), the tools of probabilistic modelling were introduced, where the specification of *what* was to be done to an image was cleanly separated from *how* it was to be done. Horn pioneered the idea of image-processing as optimisation, where the *what* step states, in the language of probability, a measure of goodness of a certain output or interpretation. Then the *how* step applies that principle by means of an algorithm that constructs, often in a series of operations, an output that most nearly satisfies the probabilistic measure. The background-removal algorithm of Figure 7.7 is of this type.

Algorithms of this sort have been very successful, but they do rely on the insight of the inventor for the right probabilistic model. The more complex the vision problem, the more difficult it is to come up with the probabilistic model. Particularly the problem of *classification* – identifying objects in images – has not proved susceptible to this approach. Inventing models or rules to capture the variability in appearance of one type of object, and contrast that with other objects, has proved too hard, for the most part. If it is not practical to characterise objects with models, what other approach could there be? The answer, of course, is

FIGURE 7.8 Object recognition. The ability to label objects and textures in an image is generated by learning from example images [26] – the illustration shows automatic tagging with a vocabulary of 20 labels.

learning by example, since this is how humans gain the ability to identify objects visually.

One breakthrough in learning objects from examples, at the turn of the millennium, was the face detector from Mitsubishi's research laboratories, mentioned earlier. Their scientists showed [15] how sets of example faces and non-faces could be used to train an algorithm to recognise a face, fast and effectively. Later, related learning algorithms were used to recognise a wider variety of objects as, for example, in Figure 7.8. Learning by example was also used to recognise the 30 or so parts of the body labelled by the Kinect 3D sensor, as in Figure 7.2.

The Rise of Deep Networks

In the current millennium, research in image classification has grown rapidly, and one particularly widespread standard test – the ImageNet challenge [11] – involves 1,000 classes of object with about 1,000 example images for each. A great deal of effort was applied to improving performance in this challenge with a variety of learning algorithms and representations of images, and around 2010–2011 performance had saturated at an error rate of about 27 per cent, as Figure 7.9 shows. Then in 2012, a further breakthrough occurred with the application of so-called *deep networks*, which, in one step, reduced the error rates dramatically [12].

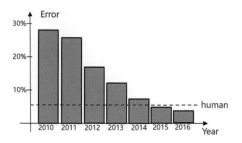

FIGURE 7.9 Reaching human performance in image recognition. The ability to label objects and textures in an image, by learning from example images [27].

In successive years, further advances in deep networks continued to push down the error rate, accumulating a drop by a factor of eight or so over just five years. In 2015, deep networks surpassed even human performance on ImageNet. That is not to say that machines now see better than humans; but, on this particular test of vision, machines are now ahead. This extraordinary development has transformed the field to the extent that now, just seven years after their introduction, nearly all published work in computer vision uses deep networks. Thanks also to accompanying advances in software tools such as Tensorflow [27], and in hardware, particularly the re-use of the graphics processor unit (GPU) for learning, deep networks have developed rapidly from obscure innovation to off-the-shelf tool, on the desk of every machine vision engineer.

Three Outstanding Challenges for Seeing Machines

The advances of 50 years or so of research in machine vision have brought about a transformation, such that machine vision is now a mainstream commercial technology. However, machines still do not see anything like as well as humans, notwithstanding a few competition wins by the machine, such as in the ImageNet challenge. There is a great deal still to be done. For instance, the area of combining image processing with language analysis is largely wide open, and is where I believe some of the major developments will happen over the next decade. To conclude this chapter, I will briefly outline three particular challenges which are currently puzzling researchers, and where more progress needs to be made.

Adversarial Attacks

Despite its striking successes, deep learning exhibits some surprising deficits. One strange behaviour was uncovered soon after deep networks began to make inroads into error rates for ImageNet. So-called *adversarial attacks*, in which small perturbations of an image, imperceptible to the human eye, would flip the class label on that image from one object class to another, were discovered [28]. In fact, it proved possible to generate perturbations of images of objects as different as a truck, a monument, and a furry dog so that, in each case, a given network would classify the image as an ostrich. Remember that the perturbations are imperceptible to the eye, so a human onlooker would continue to see a truck, a monument, and a dog, respectively. Yet the deep network would see, in each case, an ostrich.

Now this is not quite as bad as it may seem, in that it is not just any small perturbation that has this disturbing effect, but particular, highly tailored perturbations. They can be generated only by an attacker who, crucially, knows the whole design and all the settings of the particular deep network. This is known as a *white-box* attack, because it is possible only if all the details of the network are revealed. An obvious defence against this attack, then, is to keep the network details secret. Nonetheless, it is surprising and disturbing that such an attack is possible, and suggests that network architectures are frail in some respects. Despite many attempts to understand the phenomenon, adversarial attacks and how to eliminate them remain a mystery at the time of writing.

Perhaps we should not expect that adversarial vulnerability could be entirely eliminated. After all the human vision system, powerful as it is, is also susceptible to being fooled by what we know as visual illusions [22].

Few-Shot Learning

Another big challenge for machine learning in general, and for machine vision in particular, is how to learn faster. We are in the era of big data, and machines can learn effectively from copious supplies of data. This is both a bug and a feature. It's a feature because the worldwide web has brought us many kinds of data on a large scale and this abundance is now powering AI. It's a bug, because AI's learning algorithms can learn *only* when they have big data. To learn about the appearance of common objects – say cars – all is well, because there are millions of images of cars on the web, and it is usually fairly clear from context and from captions that they are indeed cars – the data are *labelled*, at least partly. So, for universal concepts learning with big data can be done. But consider the learning profile of a two- or three-year-old child. Having once learned about cars, how many images must the child now see, to learn what a truck is? Perhaps a couple, but certainly not the hundreds or thousands of images that modern machine learning requires. Psychological experiments [29] have formalised this ability of humans to learn from remarkably few examples. This is a challenge in principle for machine vision. And it is a practically important challenge in domains where data are scarce – for example, a robot exploring visually an unfamiliar environment and needing to react to it rapidly.

Safety-Critical Machines

AI is now deployed widely in industry and is essential to the vibrancy of major Internet businesses. As with any new technology, it makes sense to deploy it first in applications where the penalty for error is not too great. Following practice in various areas of engineering, we could measure the level of reliability required by an application, or delivered by a particular technology, in 9s. That is to say, using a logarithmic scale for reliability where three 9s means 99.9 per cent reliability, and so on (Figure 7.10).

On that scale, starting at the most forgiving end of the spectrum, the AI in movie recommendation might demand just one 9 reliability – if you recommend the wrong movie, really nothing very bad happens. In fact, making a good recommendation 9 times out of 10 would probably make quite a satisfied consumer. Next in line could be medical imaging, for example scanning the brain for a possible tumour, where the risks in the subsequent treatment are quite high, and the degree of agreement between expert practitioners is limited, so two 9s of reliability would likely be sufficient. Next comes face detection, where three 9s of reliability is attainable by the best face detectors (trained by learning from big data) [30]. Then automated driver assistance (ADAS) systems need to be a good deal more reliable – a driver will soon lose confidence in a car that jams on the brakes for the ghost of a jaywalker more than, say, once in

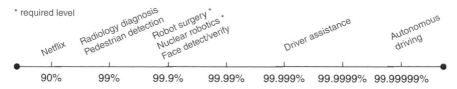

FIGURE 7.10 Safety critical technology. Appropriate levels of risk are suggested for various applications of AI. Autonomous driving is pushing AI to a new level on this spectrum.

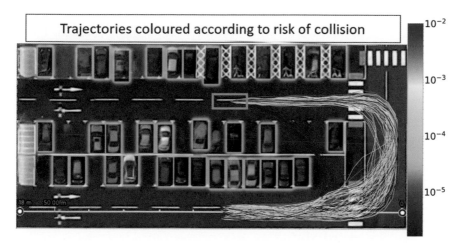

FIGURE 7.11 Computing the risk of collision for an autonomous car. In this aerial view of a car park, the probability of a collision somewhere along the path is computed for a cluster of alternative paths, for an autonomous vehicle aiming for a particular goal location. By kind permission of the FiveAI company.

1,000 miles. Assuming the ADAS makes 100 significant decisions per mile, then the necessary level of reliability would be five 9s.

Finally, entirely automated driving is now under development and, on the basis of human accident rates, the reliability of decision-making needs to reach at least seven 9s. Currently no AI system has come close to this level of demand for containment of risk. Figure 7.11 is an illustration of an experimental algorithm [31] that aims to pool probabilistic estimates of obstacle location from the sensors on an experimental, autonomous car, to estimate these very low risks of collision for a variety of proposed paths.

Conclusion

The period of 50 or so years over which computer vision has advanced from initial optimistic steps to machines that really see has been a roller coaster ride of advances interleaved with disappointments. But, over the last decade or so, we have been able to celebrate the invention and development of seeing machines that are convincing, and work at commercial scale. They are complex machines, but three important principles on which they rely are as follows: handling of the inherent uncertainty of images by computing with probabilities; learning from examples; and the use of deep neural networks. However, there are challenges ahead. Three have been picked out in particular: countering adversarial attacks on neural networks; learning faster by making better use of data; and achieving safety critical performance. I am optimistic about the progress that may be achieved on these fronts over the next decade.

References

[1] *The Economist.* GrAIt expectations. *The Economist,* 28 March 2018.

[2] McCarthy, J., Minsky, M. L., Rochester, N., and Shannon, C. E. (2006). A proposal for the Dartmouth Summer Research Project on Artificial Intelligence. *AI Mag.* 2006; 27(4): 12–14.

[3] Roberts, L. G. *Machine Perception of Three-Dimensional Solids.* PhD thesis, MIT, 1963.

[4] Papert, S. A. *The Summer Vision Project.* MIT AI Memo 100. 1966.

[5] Ambler, A. P., Barrow, H. G., Brown, C. M., Burstall, R. H., and Popplestone, R. J. A versatile computer-controlled assembly system. In *Proc. IJCAI '73: Proceedings of the 3rd International Joint Conference on Artificial Intelligence.* San Francisco, CA: Morgan Kauffmann, 1973; 298–307.

[6] Horn, B. K. P. *Robot Vision.* Cambridge, MA: MIT Press, 1986.

[7] Dickmanns, E. D., and Graefe, V. Applications of dynamic monocular machine vision. *Machine Vision Applic.* 1988; 1(4): 241–261.

[8] Thrun, S., Montemerlo, M., Dahlkamp, H., Stavens, D., Aron, A. et al. Stanley: The robot that won the DARPA Grand Challenge. *J. Field Robotics* 2006; 23(9): 661–692.

[9] Blake, A., and Isard, M. *Active Contours.* London: Springer, 1998.

[10] Cootes, T. F., Taylor, C. J., Cooper, D. H., and Graham, J. Active shape models – their training and application. *Comp. Vision Image Understanding* 1995; 61(1): 38–59.

[11] Deng, J., Dong, W., Socher, R., Li, L.-J., Li, K., and Li, F.-F. ImageNet: A large-scale hierarchical image database. In *IEEE Conference on Computational Vision and Pattern Recognition*. New York, NY: IEEE Publishing, 2009; 248–255.

[12] Krizhevsky, A., Sutskever, I., and Hinton, G. E. ImageNet classification with deep convolutional neural networks. *Adv. Neural Information Processing Syst.* 2012; 25(2): 1097–1105.

[13] He, K., Zhang, X., Ren, S., and Sun, J. Deep residual learning for image recognition. In *IEEE Conference on Computer Vision and Pattern Recognition*. New York, NY: IEEE Publishing, 2016; 770–778.

[14] Shotton, J., Fitzgibbon, A. W., Cook, M., Sharp, T., Finocchio, M. et al. Real-time human pose recognition in parts from a single depth image. *Commun. ACM* 2011; 56(1): 116–124.

[15] Viola, P., and Jones, M. Rapid object detection using a boosted cascade of simple features. In *IEEE Conference on Computer Vision and Pattern Recognition*. New York, NY: IEEE Publishing, 2001; 511–518.

[16] Kossaifi, J., Tzimiropoulos, G., Todorovic, S., and Pantic, M. AFEW-VA database for valence and arousal estimation in-the-wild. *Image and Vision Comput.* 2017; 65: 23–36.

[17] Kamnitsas, K., Ledig, C., Newcombe, V. F., Simpson, J. P., Kane, A. D. et al. Efficient multi-scale 3D CNN with fully connected CRF for accurate brain lesion segmentation. *Med. Image Anal.* 2017; 36: 61–78.

[18] Zikic, D., Ioannou, Y., Brown, M., and Criminisi, A. Segmentation of brain tumor tissues with convolutional neural networks. In *MICCAI Workshop on Multimodal Brain Tumor Segmentation Challenge (BRATS)*. New York, NY: Springer, 2014; 36–39.

[19] Mukherjee, S. (2017). A.I. versus M.D.: What happens when diagnosis is automated? *New Yorker*, 27 March 27 2017.

[20] Gavrila, D. M. Pedestrian detection from a moving vehicle. *IEEE Trans. Pattern Recognition Machine Intell.* 2000; 31(12): 2179–2195.

[21] Fleetwood, J. Public health, ethics, and autonomous vehicles. *Am. J. Public Health* 2017; 107: 532–537.

[22] Frisby, J. P. *Seeing: Illusion, Brain and Mind.* Oxford: Oxford University Press, 1979.

[23] Gregory, R. L. *Eye and Brain: The Psychology of Seeing.* Princeton, NJ: Princeton University Press, 1966.

[24] Besag, J. Spatial interaction and the statistical analysis of lattice systems. *J. Roy. Statist. Soc. B* 1974; 36(2): 192–225.

[25] Rother, C., Kolmogorov, V., and Blake, A. 'GrabCut': Interactive foreground extraction using iterated graph cuts. *ACM Trans. Graphics* 2004; 23(3): 309–314.

[26] Shotton, J., Winn, J., Rother, C., and Criminisi, A. *TextonBoost*: Joint appearance, shape and context modeling for multi-class object recognition and segmentation. In Leonardis, A., Bischof, H., and Pinz, A., eds. *Computer Vision – ECCV 2006: 9th European Conference on Computer Vision, Graz, Austria, May 7–13, 2006. Proceedings, Part I.* Berlin: Springer, 2006; 1–15.

[27] Abadi, M., Barham, P., Chen, J., Chen, Z., Davis, A. et al. TensorFlow: A system for large-scale machine learning. In *Proc. 12th USENIX Symposium on Operating Systems Design and Implementation (OSDI '16).* 2016; 265–283.

[28] Szegedy, C., Zaremba, W., Sutskever, I., Bruna, J., Erhan, D. et al. Intriguing properties of neural networks. 2013; arXiv preprint arXiv:1312.6199.

[29] Salakhutdinov, R., Tenenbaum, J., and Torralba, A. One-shot learning with a hierarchical nonparametric Bayesian model. *J. Machine Learning Res.* 2012; 27: 195–206.

[30] Schroff, F., Kalenichenko, D., and Philbin, J. FaceNet: A unified embedding for face recognition and clustering. In *2015 IEEE Conference on Computer Vision and Pattern Recognition.* New York, NY: IEEE Publishing, 2015; 815–823.

[31] Blake, A., Bordallo, A., Hawasly, M., Penkov, S., Ramamoorthy, S., and Silva, A. Efficient Computation of Collision Probabilities for Safe Motion Planning. 2018; Arxiv 1804.05384.

Index

Index